First Workshop on Dynamic Adversarial Data Collection (DADC 2022)

Online
14 July 2022

ISBN: 978-1-7138-5627-6

Printed from e-media with permission by:

Curran Associates, Inc.
57 Morehouse Lane
Red Hook, NY 12571

Some format issues inherent in the e-media version may also appear in this print version.

Printed with permission by Curran Associates, Inc. (2022)

For permission requests, please contact the Association for Computational Linguistics at the address below.

Association for Computational Linguistics
209 N. Eighth Street
Stroudsburg, Pennsylvania 18360

Phone: 1-570-476-8006
Fax: 1-570-476-0860

acl@aclweb.org

Additional copies of this publication are available from:

Curran Associates, Inc.
57 Morehouse Lane
Red Hook, NY 12571 USA
Phone: 845-758-0400
Fax: 845-758-2633
Email: curran@proceedings.com
Web: www.proceedings.com

DADC 2022

The First Workshop on Dynamic Adversarial Data Collection (DADC)

Proceedings of the Workshop

July 14, 2022

The DADC organizers gratefully acknowledge the support from the following sponsors.

Platinum

ML
• Commons

Gold

Meta

Order copies of this and other ACL proceedings from:

Association for Computational Linguistics (ACL)
209 N. Eighth Street
Stroudsburg, PA 18360
USA
Tel: +1-570-476-8006
Fax: +1-570-476-0860
acl@aclweb.org

Preface

This volume contains papers from the First Workshop on Dynamic Adversarial Data Collection (DADC), held at NAACL 2022.
Dynamic Adversarial Data Collection (DADC) has been gaining traction in the community as a promising approach to improving data collection practices, model evaluation and performance. DADC allows us to collect human-written data dynamically with models in the loop. Humans can be tasked with finding adversarial examples that fool current state-of-the-art models (SOTA), for example, or they can cooperate with models to find interesting examples. This offers two benefits: it allows us to gauge how good contemporary SOTA methods really are; and it yields data that may be used to train even stronger models by specifically targeting their current weaknesses.
The first workshop on DADC and corresponding shared task focus on three currently under-explored themes: i) understanding how humans can be incentivized to creatively identify and target model weaknesses to increase their chances of fooling the model; ii) how humans and machines can cooperate to produce the most useful data; and iii) how the interaction between humans and machines can further drive performance improvements, both from the perspectives of traditional evaluation metrics as well as those of robustness and fairness.

Organizing Committee

General Chairs

Max Bartolo, University College London
Hannah Rose Kirk, University of Oxford
Pedro Rodriguez, FAIR Labs, Seattle
Katerina Margatina, University of Sheffield
Tristan Thrush, Hugging Face
Robin Jia, University of Southern California

Advisory Committee

Pontus Stenetorp, University College London
Adina Williams, FAIR, NYC
Douwe Kiela, Hugging Face

Program Committee

Program Committee

Giorgos Vernikos, EPFL & HEIG-VD
John P. Lalor, University of Notre Dame
Maharshi Gor, University of Maryland, College Park
Pasquale Minervini, University College London
Paul Rottger, University of Oxford
Shi Feng, University of Maryland, College Park
Unso Eun Seo Jo, Stanford University & HuggingFace

Invited Speakers

Anna Rogers, University of Copenhagen
Sam Bowman, New York University
Jordan Boyd-Graber, University of Maryland
Lora Aroyo, Google
Tongshuang Wu, Carnegie Mellon University

Keynote Talk: What kinds of questions have we been asking? A taxonomy for QA/RC benchmarks

Anna Rogers
University of Copenhagen

Abstract: This talk provides an overview of the current landscape of resources for Question Answering and Reading comprehension, highlighting the current lacunae for future work. I will also present a new taxonomy of "skills" targeted by QA/RC datasets and discuss various ways in which questions may be unanswerable.

Bio: Anna Rogers is an Assistant Professor in the Center for Social Data Science at the University of Copenhagen. She is also a visiting researcher with the RIKEN Center for Computational Science. Anna's main research area is Natural Language Processing, in particular model analysis and evaluation of natural language understanding systems.

Keynote Talk: Why Adversarially-Collected Test Sets Don't Work as Benchmarks

Sam Bowman
New York University

Abstract: Dynamic and/or adversarial data collection can be quite useful as a way of collecting training data for machine-learning models, identifying the conditions under which these models fail, and conducting online head-to-head comparisons between models. However, it is essentially impossible to use these practices to build usable static benchmark datasets for use in evaluating or comparing future new models. I defend this point using a mix of conceptual and empirical points, focusing on the claims (i) that adversarial data collection can skew the distribution of phenomena such as to make it unrepresentative of the intended task, and (ii) that adversarial data collection can arbitrarily shift the rankings of models on its resulting test sets to disfavor systems that are qualitatively similar to the current state of the art.

Bio: Sam Bowman is an Assistant Professor at New York University and a Visiting Researcher (Sabbatical) at Anthropic. His research interests include the study of artificial neural network models for natural language understanding, with a focus on building high-quality training and evaluation data, applying these models to scientific questions in syntax and semantics, and contributing to work on language model alignment and control.

Keynote Talk: Incentives for Experts to Create Adversarial QA and Fact-Checking Examples

Jordan Boyd-Graber
University of Maryland

Abstract: I'll discuss two examples of our work putting experienced writers in front of a retrieval-driven adversarial authoring system: question writing and fact-checking. For question answering, we develop a retrieval-based adversarial authoring platform and create incentives to get people to use our system in the first place, write interesting questions humans can answer, and challenge a QA system. While the best humans lose to computer QA systems on normal questions, computers struggle to answer our adversarial questions. We then turn to fact checking, creating a new game (Fool Me Twice) to solicit difficult-to-verify claims—that can be either true or false—and to test how difficult the claims are both for humans and computers. We argue that the focus on retrieval is important for knowledge-based adversarial examples because it highlights diverse information, prevents frustration in authors, and takes advantage of users' expertise.

Bio: Jordan Boyd-Graber is an Associate Professor in the University of Maryland Computer Science Department (tenure home), Institute of Advanced Computer Studies, iSchool, and Language Science Center. Previously, he was an Assistant Professor at Colorado's Department of Computer Science (tenure granted in 2017). Jordan's research focuses on making machine learning more useful, more interpretable, and able to learn and interact from humans.

Keynote Talk: Data Excellence: Better Data for Better AI

Lora Aroyo
Google

Abstract: The efficacy of machine learning (ML) models depends on both algorithms and data. Training data defines what we want our models to learn, and testing data provides the means by which their empirical progress is measured. Benchmark datasets define the entire world within which models exist and operate, yet research continues to focus on critiquing and improving the algorithmic aspect of the models rather than critiquing and improving the data with which our models operate. If "data is the new oil," we are still missing work on the refineries by which the data itself could be optimized for more effective use. In this talk, I will discuss data excellence and lessons learned from software engineering to achieve the scare and rigor in assessing data quality.

Bio: Lora Aroyo is Research Scientist at Google Research, NYC, where she works on research for Data Excellence by specifically focussing on metrics and strategies to measure quality of human-labeled data in a reliable and transparent way. Lora is an active member of the Human Computation, User Modeling and Semantic Web communities. She is president of the User Modeling community UM Inc, which serves as a steering committee for the ACM Conference Series "User Modeling, Adaptation and Personalization" (UMAP) sponsored by SIGCHI and SIGWEB. She is also a member of the ACM SIGCHI conferences board. Prior to joining Google, Lora was a computer science professor at the VU University Amsterdam.

Keynote Talk: Model-in-the-loop Data Collection: What Roles does the Model Play?

Tongshuang Wu
Carnegie Mellon University

Abstract: Assistive models have been shown useful for supporting humans in creating challenging datasets, but how exactly do they help? In this talk, I will discuss different roles of assistive models in counterfactual data collection (i.e., perturbing existing text inputs to gain insight into task model decision boundaries), and the characteristics associated with these roles. I will use three examples (CheckList, Polyjuice, Tailor) to demonstrate how our objectives shift when we perturb texts for evaluation, explanation, and improvement, and how that change the corresponding assistive models from enhancing human goals (requiring model controllability) to competing with human bias (requiring careful data reranking). I will conclude by exploring additional roles that these models can play to become more effective.

Bio: Sherry Tongshuang Wu is an Assistant Professor at the Human Computer Interaction Institute at Carnegie Mellon University (CMU HCII), holding a courtesy appointment at the Language Technolgoy Institute (CMU LTI). Sherry's research lies at the intersection of Human-Computer Interaction (HCI) and Natural Language Processing (NLP). She aims to understand and support people coping with imperfect AI models, both when the model is under active development, and after it is deployed for end users.

Table of Contents

Program

Thursday, July 14, 2022

09:00 - 09:10	*Opening Remarks*
09:10 - 09:25	*Collaborative Progress: ML Commons Introduction*
09:25 - 10:00	*Invited Talk 1: Anna Rogers*
10:00 - 10:35	*Invited Talk 2: Jordan Boyd-Graber*
10:35 - 10:50	*Break*
10:50 - 11:10	*Best Paper Talk:*
	Overconfidence in the Face of Ambiguity with Adversarial Data Margaret Li and Julian Michael
11:10 - 11:45	*Invited Talk 3: Sam Bowman*
11:45 - 12:20	*Invited Talk 4: Lora Aroyo*
12:20 - 13:20	*Lunch*
13:20 - 13:55	*Invited Talk 5: Sherry Tongshuang Wu*
13:55 - 14:55	*Panel: The Future of Data Collection*
14:55 - 15:10	*Break*
15:10 - 15:20	*Introduction to the DADC Shared Task: Max Bartolo*
15:20 - 15:40	*Shared Task Winners' Presentations*
15:40 - 16:55	*Poster Session*

Thursday, July 14, 2022 (continued)

Resilience of Named Entity Recognition Models under Adversarial Attack
Sudeshna Das and Jiaul Paik

GreaseVision: Rewriting the Rules of the Interface
Siddhartha Datta, Konrad Kollnig and Nigel Shadbolt

Posthoc Verification and the Fallibility of the Ground Truth
Yifan Ding, Nicholas Botzer and Tim Weninger

Overconfidence in the Face of Ambiguity with Adversarial Data
Margaret Li and Julian Michael

longhorns at DADC 2022: How many linguists does it take to fool a Question Answering model? A systematic approach to adversarial attacks.
Venelin Kovatchev, Trina Chatterjee, Venkata S Govindarajan, Jifan Chen, Eunsol Choi, Gabriella Chronis, Anubrata Das, Katrin Erk, Matthew Lease, Junyi Jessy Li, Yating Wu and Kyle Mahowald

Collecting high-quality adversarial data for machine reading comprehension tasks with humans and models in the loop
Damian Y. Romero Diaz, Magdalena Anioł and John Culnan

Generalized Quantifiers as a Source of Error in Multilingual NLU Benchmarks
Ruixiang Cui, Daniel Hershcovich and Anders Søgaard

Adversarially Constructed Evaluation Sets Are More Challenging, but May Not Be Fair
Jason Phang, Angelica Chen, William Huang and Samuel R. Bowman

16:55 - 17:00 *Closing Session*

Resilience of Named Entity Recognition Models Under Adversarial Attack

Sudeshna Das
Indian Institute of Technology
Kharagpur
sudeshna.das@iitkgp.ac.in

Jiaul H Paik
Indian Institute of Technology
Kharagpur
jiaul@cet.iitkgp.ac.in

Abstract

Named entity recognition (NER) is a popular language processing task with wide applications. Progress in NER has been noteworthy, as evidenced by the F1 scores obtained on standard datasets. In practice, however, the end-user uses an NER model on their dataset out-of-the-box, on text that may not be pristine. In this paper we present four model-agnostic adversarial attacks to gauge the resilience of NER models in such scenarios. Our experiments on four state-of-the-art NER methods with five English datasets suggest that the NER models are over-reliant on case information and do not utilise contextual information well. As such, they are highly susceptible to adversarial attacks based on these features.

1 Introduction

Named entity recognition (NER) is a popular language processing task that involves identifying and classifying named entities in text (Mayhew et al., 2020). Progress in NER has been rapid and noteworthy, especially in the current age of deep learning (Li et al., 2020). The general impetus in deep learning-based NER has been to develop models that incorporate context better (Akbik et al., 2018; Devlin et al., 2019; Manning et al., 2014) and are resilient to noise such as inconsistencies in case information (Mayhew et al., 2019; Bodapati et al., 2019; Mayhew et al., 2020). There has, however, been modest focus on determining the extent to which state-of-the-art NER models succeed in doing so. Identifying the weaknesses of NER models can help drive focused work to ameliorate them and move NER beyond marginal improvements in F1 scores (Stanislawek et al., 2019).

Adversarial attacks designed for NLP models largely focus on classification tasks (Wallace et al., 2019; Ren et al., 2019; Jia et al., 2019; Wallace et al., 2019; Papernot et al., 2016). Many existing studies work with vector representations (Ebrahimi et al., 2018; Zhao et al., 2018), which are not intuitively interpretable by humans. Such methods require white-box access to the models (Ren et al., 2019). The additional requirement of human intervention to adjudge the quality of adversarial samples generated may also be involved (Alzantot et al., 2018).

Adversarial NER has broadly seen two types of approaches: (a) adversarial training, and (b) adversarial evaluation. Adversarial training of NER models involves introducing small perturbations in the training data to make models robust (Bekoulis et al., 2018). Such perturbations are introduced in the text representation level (Wang et al., 2020; Bai et al., 2020; Huang et al., 2022). The adversarial evaluation of NER models, on the other hand, involves benchmarking the models on synthetically generated data (Lin et al., 2021; Simoncini and Spanakis, 2021). We follow the latter line of investigation.

We present four model-agnostic adversarial attacks targeted at NER models. Our task-specific approach allows us to generate natural language adversaries that work with pre-trained models and are easily interpretable by humans. In principle, our work is similar to the label-preserving substitutions explored by Ren et al. (2019) and the word-substitution methods explored by Alzantot et al. (2018), although they do not evaluate their methods on NER. Generating adversarial data for evaluating NER models is explored by Simoncini and Spanakis (2021) using BERT to replace and/or add non-named entity tokens to text. Lin et al. (2021) also use pre-trained BERT to generate context-level adversarial attacks to evaluate NER models. In contrast to their work, we use simple rule-based methods for generating adversarial data. Our method has the advantage of not requiring re-training or fine-tuning of pre-trained models.

The datasets and models we use are all openly available, aiding reproducibility.[1] Further, our ex-

[1] https://github.com/das-sudeshna/adversarial-ner

Proceedings of the First Workshop on Dynamic Adversarial Data Collection, pages 1 - 6
July 14, 2022 ©2022 Association for Computational Linguistics

	CoNLL	WIKI	GMB	FIRE	IEER
# LOCATION	1668	1014	59255	2626	878
# PERSON	1617	934	18970	2725	1504
# ORGANIZATION	1661	898	22662	893	939

Table 1: Data description: frequency of named entities.

periments do not require white-box access to the models.

2 Data

We use five openly available general domain datasets that contain the *enamex* classes (LOCATION, PERSON, & ORGANIZATION) (Nadeau and Sekine, 2007), for this study.

CoNLL-2003 The CoNLL-2003 (CoNLL) dataset consists of news articles from the Reuters Corpus (Tjong Kim Sang and De Meulder, 2003). In keeping with the standard evaluation schemes, we report results only on the test split of the dataset.

WikiGold The WikiGold dataset (WIKI) comprises of manually annotated English Wikipedia articles (Balasuriya et al., 2009).

FIRE NER 2013 The English dataset (FIRE) from the *NER for Indian Languages* task at FIRE 2013 comprises of text crawled from Indian websites as well as Wikipedia articles.

NIST IE-ER 1999 IEER refers to the gold standard NEWSWIRE development test data for the NIST 1999 IE-ER Evaluation available with NLTK (Steven Bird and Klein, 2009).

GMB 2.2 The Groningen Meaning Bank 2.2 dataset comprises of public domain texts that include news articles, stories, jokes, and transcripts. NLP tools are used to provide a preliminary annotation which is then updated by a combination of human experts, NLP tools, and crowd-sourcing to yield a silver-standard corpus (Bos et al., 2017).

3 Methods

We use four named entity recognizers for our experiments, all of which are open-source. Of these, spaCy is the current state-of-the-art in terms of document processing speed (Choi et al., 2015) and Flair is near the current state-of-the-art.[2]

Flair NER The Flair named entity recognizer is based on neural character embeddings. It uses contextual neural string embeddings that are obtained by pre-training on large, unlabelled corpora. Every sentence is represented in the form of string embeddings which are then stacked with pre-computed uncased GloVe embeddings, before being passed through a BiLSTM-CRF architecture that generates labels for each word (Akbik et al., 2018).

spaCy NER spaCy's named entity recognizer employs a transition-based entity recognition methodology where state changes are triggered by actions. It uses trigram CNNs with residual connections that transform context-independent vectors into context-sensitive vectors (Honnibal, 2016).

CoreNLP NER CoreNLP NER (Manning et al., 2014) is based on linear chain Conditional Random Field (CRF) sequence models of arbitrary order (Finkel et al., 2005). For our experiments, we use the caseless model that ignores capitalization as well as the Truecase annotator that attempts to rectify incorrect casing, in addition to the default model.

DeepPavlov NER DeepPavlov's named entity recognition model uses the English cased model of BERT with 12 layers, 768 hidden nodes, 12 attention heads, and 110M parameters (Devlin et al., 2019). The first sub-word representation of each word is passed through a dense layer to generate labels (Burtsev et al., 2018).

Original:	My sister Alice (PERSON) lives in London (LOCATION).
Case	
Ablation:	my sister alice (PERSON) lives in london (LOCATION).
Aberration:	My sister alice (PERSON) Lives in London (LOCATION).
Context	
Perturbation:	My sister London (PERSON) lives in Alice (LOCATION).
Alteration:	My sister lives Alice (PERSON*) London (LOCATION*) in.

Figure 1: Dataset variants. Class* denotes named entities that should desirably be misclassified from their context.

4 Adversarial Attacks

In this section we describe the design of two broad types of adversarial attacks on NER models.

[2]The F1 score of the current state-of-the-art model is 0.935. (Flair's F1 score is 0.931.) Since a pre-trained model is not publicly available, we choose not to include it in our experiments. We strongly believe that this does not affect the conclusions of our work.

4.1 Case-based Adversarial Attacks

Case is one of the strongest indicators of named entities in English (Mayhew et al., 2020) and it is well known that case affects the performance of NER models (Mayhew et al., 2019; Bodapati et al., 2019). We formulate two adversarial attacks that emulate data where (i) case information may be unavailable, such as informal texts, and (ii) case information is unreliable, such as text extracted from PDF or OCR-ed documents.

4.1.1 Case Ablation

In case ablation, we drop the case information while keeping the rest of the text intact. The case-ablated named entities attempt to fool the NER models into misclassifying them as non-entities. This allows us to quantify what percentage of the correctly identified named entities rely completely on case information.

4.1.2 Case Aberration

In this setup, we randomly capitalise N percent of the tokens in each dataset, where N is the percentage of actual named entity tokens in the corresponding original text. The randomly capitalised tokens attempt to fool the model into marking them as named entities. We choose N rather than an arbitrary value in order to maintain the distribution of capitalised and lowercase tokens in the datasets.

4.2 Context-based Adversarial Attacks

The surrounding text of a named entity is arguably the most useful feature in identifying named entities. All the NER models we evaluate attempt to capture context to leverage this information. We formulate two adversarial attacks that attempt to determine how well such information is captured by these models.

4.2.1 Context Perturbation

We create local perturbations for named entities. That is, we change the immediately surrounding text of the named entities while retaining syntactic structure and a semblance of semantics. To achieve this, we replace named entities of each class by named entities of the other two classes, with an equal probability. The local context of a named entity attempts to fool the NER model into classifying it incorrectly. This attack is similar in nature to the data augmentation procedure used by Lin et al. (2021). However, they restrict named entity substitutions within the same entity class. Since we carry out inter-class entity substitutions, we posit that our method is better able to detect when NER models rely on memorising named entity tokens.

4.2.2 Context Alteration

We alter the context of named entities on a global scale. To achieve this, we randomly select named entities with equal probability and place them in random locations in the text. In almost all cases, the text becomes grammatically incorrect, as is illustrated in Table 1. Thus, neither semantics nor syntactic rules are maintained, effectively altering the global contextual frame of named entities. In this case, it is *desirable* for models to misclassify named entities. That is, we consider a model to be better if it is *susceptible* to this attack. This is based on our hypothesis that a model that captures context better should perform *worse* when the context is meaningless.

5 Evaluation

We follow the CoNLL-2003 Shared Task guidelines to report the F1 scores (Tjong Kim Sang and De Meulder, 2003). Compatible classes are clubbed with the closest enamex class (such as, GPE (Geo-political entity) is clubbed with LOCATION for spaCy, BERT, and the GMB dataset). The class labels present in different datasets/produced by different models do not always have a close one-to-one correspondence to the class labels in other datasets/produced by other models. Thus, non-enamex entities are considered to be non-entities to provide a fair comparison across datasets and models. NER models and datasets also differ in their tagging schemes. Since it is not possible to map IO tags to IOB or IOBES, and IOB tags to IOBES (Cho et al., 2013), we map all tags into the IO scheme. The mapping of compatible entity classes and tagging schemes causes our evaluation results to differ from the officially reported scores of these NER models.

Model	CoNLL	WIKI	GMB	FIRE	IEER
Flair	**0.92**	**0.92**	**0.93**	**0.93**	**0.93**
spaCy	0.85	0.89	0.91	0.90	0.90
CoreNLP-s	0.86	0.89	0.92	0.88	0.90
DeepPavlov	0.83	0.90	0.90	0.90	0.92

Table 2: F1 scores on original datasets.

6 Results and Analysis

Table 2 shows the F1 scores of the models on the original dataset. This gives us the benchmark against which we compare the performance for the different data variants.

Model	CoNLL	WIKI	GMB	FIRE	IEER
Flair	**0.37** (-52.03%)	0.16 (-74.24%)	0.35 (-54.48%)	0.18 (-73.66%)	0.14 (-77.97%)
spaCy	0.14 (-68.64%)	0.13 (-74.06%)	0.20 (-68.32%)	0.15 (-72.97%)	0.14 (-75.02%)
CoreNLP-s	0.19 (-62.81%)	0.11 (-75.73%)	0.28 (-61.76%)	0.16 (-69.57%)	0.12 (-77.11%)
CoreNLP-c	0.32 **(-47.72%)**	**0.23 (-62.75%)**	**0.42 (-46.74%)**	**0.30 (-53.83%)**	**0.28 (-59.58%)**
CoreNLP-t	0.20 (-62.69%)	0.11 (-75.96%)	0.28 (-61.76%)	0.16 (-70.03%)	0.11 (-77.78%)
DeepPavlov	0.25 (-52.45%)	0.15 (-73.30%)	0.16 (-72.42%)	0.23 (-63.91%)	0.13 (-77.77%)

Table 3: F1 scores on case ablated datasets. High F1 score and low percentage drops are desirable.

Model	CoNLL	WIKI	GMB	FIRE	IEER
Flair	**0.39** (-50.19%)	0.21 (-69.00%)	0.36 (-53.62%)	0.23 (-68.50%)	0.21 (-71.33%)
spaCy	0.20 (-61.21%)	0.19 (-67.43%)	0.26 (-61.70%)	0.20 (-67.61%)	0.19 (-69.82%)
CoreNLP-s	0.25 (-56.38%)	0.17 (-69.19%)	0.32 (-57.08%)	0.21 (-63.75%)	0.17 (-71.34%)
CoreNLP-c	0.32 **(-47.72%)**	**0.23 (-62.75%)**	**0.42 (-46.74%)**	**0.30 (-53.83%)**	**0.28 (-59.58%)**
CoreNLP-t	0.25 (-56.38%)	0.17 (-69.19%)	0.32 (-57.08%)	0.20 (-64.55%)	0.16 (-72.12%)
DeepPavlov	0.32 (-44.46%)	0.15 (-72.29%)	0.22 (-65.51%)	0.26 (-60.23%)	0.19 (-71.59%)

Table 4: F1 scores on case aberrated datasets. High F1 score and low percentage drops are desirable.

6.1 Case ablation

We observe significantly large performance drops for every model with respect to model performance on the original datasets. This is unsurprising, as case information is an important indicator of named entities.

If we consider the CoreNLP-c scores as the upper bound (since this model is trained on caseless data and hence, reflects the ability of NER models to work on caseless data), we still notice large drops in F1 scores for the other models. This reflects the tendency of NER models to over-rely on case information. Among the cased models, we find BERT to be the better performer with Flair trailing as a close competitor. This is an interesting finding as it suggests that cased BERT is more resilient to case-based adversarial attacks than Flair, which uses uncased GloVe embeddings.

6.2 Case aberration

We observe large drops in performance for the case aberration attack. The performance for CoreNLP-t is worse than that of CoreNLP-c, which suggests that truecasing is not as effective as caseless training. Among the case-sensitive models, we find Flair outperforming other models. The performance drop for case aberration is slightly less than that for case ablation.

Model	CoNLL	WIKI	GMB	FIRE	IEER
Flair	**0.22** (-68.53%)	**0.27 (-62.05%)**	**0.23 (-67.65%)**	**0.19 (-73.05%)**	**0.18 (-74.22%)**
spaCy	0.15 (-67.49%)	0.20 (-66.96%)	0.19 (-69.74%)	0.13 (-74.71%)	0.12 (-77.03%)
CoreNLP-s	0.16 **(-67.02%)**	0.17 (-68.99%)	0.19 (-71.52%)	0.13 (-73.37%)	0.09 (-79.73%)
DeepPavlov	0.11 (-69.28%)	0.10 (-78.33%)	0.10 (-78.89%)	0.08 (-80.11%)	0.10 (-81.46%)

Table 5: F1 scores on context perturbed datasets. High F1 score and low percentage drops are desirable.

6.3 Context perturbation

Despite including mechanisms to incorporate contextual information, NER models show large performance drops under context perturbation attacks. Since an NER model is highly likely to have come across *"London"* as a LOCATION and *"Alice"* as a PERSON during training, it predicts them as such, ignoring the local context in which they appear. Despite large performance drops in general, Flair outperforms other models for all five datasets. This suggests that Flair captures local context better, likely due to the use of character embeddings.

Model	CoNLL	WIKI	GMB	FIRE	IEER
Flair	0.36 (-53.55%)	0.20 **(-69.33%)**	0.40 (-48.97%)	0.27 **(-64.31%)**	0.28 **(-63.50%)**
spaCy	0.27 (-52.94%)	0.19 (-67.20%)	0.38 (-49.02%)	0.26 (-61.03%)	0.27 (-61.09%)
CoreNLP-s	0.29 (-51.93%)	0.20 (-66.14%)	0.38 (-50.77%)	0.27 (-56.57%)	0.28 (-58.69%)
DeepPavlov	0.23 **(-55.00%)**	0.21 (-66.59%)	0.35 (-57.21%)	0.29 **(-51.13%)**	0.27 (-63.02%)

Table 6: F1 scores on context altered datasets. High percentage drops are desirable.

6.4 Context alteration

We note here that unlike the previous experiments, it is desirable to have higher percentage drops in performance for the context alteration attacks.[3] All the models show drops in performance. This hints at NER models having a tendency to learn the names themselves during training, rather than relying on the context in which the names appear. The magnitude of drops in performance is generally less than that observed for context perturbation, which suggests that NER models capture the local context of named entities better than their global context. Flair shows the largest performance drops, closely trailed by BERT.

7 Discussion

The adversarial evaluation of NLP models rely either on human-generated adversaries (Kaushik et al., 2019) or automated adversary generation with human-in-the-loop (Alzantot et al., 2018). However, it is possible to do away with human intervention for generating adversarial samples for the task of NER, as we demonstrate. Further, unlike existing work, our approach for adversarial evaluation does not require any re-training or fine-tuning of models for adversarial data creation.

The generalizability of NER models can also be evaluated with the proposed approaches. In particular, context perturbation can be used as an alternative to studying the effect of named entities that have not been seen during training (Augenstein et al., 2017) with the same label.

8 Conclusions

In this paper, we present an adversarial evaluation of four popular named-entity recognizers on five English datasets. The four model-agnostic adversarial attacks we present do not require white-box access to pre-trained NER models. Our experiments show that the popular NER models are over-reliant on the case information and under-utilise the contextual information. Since NER is a pre-requisite for a large number of NLP tasks, further work for improvement in these directions is warranted.

[3]Lower F1 scores are also desirable. However, low F1 scores can also be caused due to a model being poor generally and not specifically due to the inability to capture global context. Thus, we cannot draw concrete conclusions from the absolute F1 scores.

References

Alan Akbik, Duncan Blythe, and Roland Vollgraf. 2018. Contextual string embeddings for sequence labeling. In *Proceedings of the 27th International Conference on Computational Linguistics*, pages 1638–1649.

Moustafa Alzantot, Yash Sharma, Ahmed Elgohary, Bo-Jhang Ho, Mani Srivastava, and Kai-Wei Chang. 2018. Generating natural language adversarial examples. In *Proceedings of the Conference on Empirical Methods in Natural Language Processing*, pages 2890–2896.

Isabelle Augenstein, Leon Derczynski, and Kalina Bontcheva. 2017. Generalisation in named entity recognition: A quantitative analysis. *Computer Speech & Language*, 44:61–83.

Yuxuan Bai, Yu Wang, Bin Xia, Yun Li, and Ziye Zhu. 2020. Adversarial named entity recognition with pos label embedding. In *2020 International Joint Conference on Neural Networks*, pages 1–8. IEEE.

Dominic Balasuriya, Nicky Ringland, Joel Nothman, Tara Murphy, and James R Curran. 2009. Named entity recognition in wikipedia. In *Proceedings of the Workshop on The People's Web Meets NLP*, pages 10–18.

Giannis Bekoulis, Johannes Deleu, Thomas Demeester, and Chris Develder. 2018. Adversarial training for multi-context joint entity and relation extraction. In *Proceedings of the Conference on Empirical Methods in Natural Language Processing*, pages 2830–2836.

Sravan Bodapati, Hyokun Yun, and Yaser Al-Onaizan. 2019. Robustness to capitalization errors in named entity recognition. In *Proceedings of the 5th Workshop on Noisy User-generated Text*, pages 237–242.

Johan Bos, Valerio Basile, Kilian Evang, Noortje J Venhuizen, and Johannes Bjerva. 2017. The groningen meaning bank. In *Handbook of linguistic annotation*, pages 463–496.

Mikhail Burtsev, Alexander Seliverstov, Rafael Airapetyan, Mikhail Arkhipov, Dilyara Baymurzina, Nickolay Bushkov, Olga Gureenkova, Taras Khakhulin, Yurii Kuratov, Denis Kuznetsov, et al. 2018. Deeppavlov: Open-source library for dialogue systems. In *Proceedings of ACL 2018*, pages 122–127.

Han-Cheol Cho et al. 2013. Named entity recognition with multiple segment representations. *Information Processing & Management*, 49(4):954–965.

Jinho D Choi, Joel Tetreault, and Amanda Stent. 2015. It depends: Dependency parser comparison using a web-based evaluation tool. In *Proceedings of the 53rd Annual Meeting of the Association for Computational Linguistics and the 7th International Joint Conference on Natural Language Processing*, pages 387–396.

Jacob Devlin, Ming-Wei Chang, Kenton Lee, and Kristina Toutanova. 2019. Bert: Pre-training of deep bidirectional transformers for language understanding. In *Proceedings of the Conference of the North American Chapter of the Association for Computational Linguistics: Human Language Technologies*, pages 4171–4186.

Javid Ebrahimi, Anyi Rao, Daniel Lowd, and Dejing Dou. 2018. Hotflip: White-box adversarial examples for text classification. In *Proceedings of the 56th Annual Meeting of the Association for Computational Linguistics*, pages 31–36.

Jenny Rose Finkel, Trond Grenager, and Christopher D Manning. 2005. Incorporating non-local information into information extraction systems by gibbs sampling. In *Proceedings of the 43rd Annual Meeting of the Association for Computational Linguistics*, pages 363–370.

Matthew Honnibal. 2016. spacy.

Peixin Huang, Xiang Zhao, Minghao Hu, Yang Fang, Xinyi Li, and Weidong Xiao. 2022. Extract-select: A span selection framework for nested named entity recognition with generative adversarial training. In *Findings of the Association for Computational Linguistics*, pages 85–96.

Robin Jia, Aditi Raghunathan, Kerem Göksel, and Percy Liang. 2019. Certified robustness to adversarial word substitutions. In *Proceedings of the Conference on Empirical Methods in Natural Language Processing and the 9th International Joint Conference on Natural Language Processing*.

Divyansh Kaushik, Eduard Hovy, and Zachary Lipton. 2019. Learning the difference that makes a difference with counterfactually-augmented data. In *International Conference on Learning Representations*.

Jing Li, Aixin Sun, Jianglei Han, and Chenliang Li. 2020. A survey on deep learning for named entity recognition. *IEEE Transactions on Knowledge and Data Engineering*.

Bill Yuchen Lin, Wenyang Gao, Jun Yan, Ryan Moreno, and Xiang Ren. 2021. Rockner: A simple method to create adversarial examples for evaluating the robustness of named entity recognition models. In *Proceedings of the Conference on Empirical Methods in Natural Language Processing*, pages 3728–3737.

Christopher D Manning, Mihai Surdeanu, John Bauer, Jenny Rose Finkel, Steven Bethard, and David McClosky. 2014. The stanford corenlp natural language processing toolkit. In *Proceedings of 52nd annual meeting of the association for computational linguistics*, pages 55–60.

Stephen Mayhew, Gupta Nitish, and Dan Roth. 2020. Robust named entity recognition with truecasing pretraining. In *Proceedings of the AAAI Conference on Artificial Intelligence*, volume 34, pages 8480–8487.

Stephen Mayhew, Tatiana Tsygankova, and Dan Roth. 2019. ner and pos when nothing is capitalized. In *Proceedings of the Conference on Empirical Methods in Natural Language Processing and the 9th International Joint Conference on Natural Language Processing*, pages 6257–6262.

David Nadeau and Satoshi Sekine. 2007. A survey of named entity recognition and classification. *Lingvisticae Investigationes*, 30(1):3–26.

Nicolas Papernot, Patrick McDaniel, Ananthram Swami, and Richard Harang. 2016. Crafting adversarial input sequences for recurrent neural networks. In *IEEE Military Communications Conference*, pages 49–54. IEEE.

Shuhuai Ren, Yihe Deng, Kun He, and Wanxiang Che. 2019. Generating natural language adversarial examples through probability weighted word saliency. In *Proceedings of the 57th Annual Meeting of the Association for Computational Linguistics*, pages 1085–1097.

Walter Simoncini and Gerasimos Spanakis. 2021. Seqattack: On adversarial attacks for named entity recognition. In *Proceedings of the Conference on Empirical Methods in Natural Language Processing*, pages 308–318.

Tomasz Stanislawek, Anna Wróblewska, Alicja Wójcicka, Daniel Ziembicki, and Przemyslaw Biecek. 2019. Named entity recognition - is there a glass ceiling? In *Proceedings of the 23rd Conference on Computational Natural Language Learning*, pages 624–633.

Edward Loper Steven Bird and Ewan Klein. 2009. *Natural Language Processing with Python.*

Erik F Tjong Kim Sang and Fien De Meulder. 2003. Introduction to the conll-2003 shared task: language-independent named entity recognition. In *Proceedings of the 7th conference on Natural language learning at HLT-NAACL 2003*, pages 142–147.

Eric Wallace, Shi Feng, Nikhil Kandpal, Matt Gardner, and Sameer Singh. 2019. Universal adversarial triggers for attacking and analyzing nlp. In *Proceedings of the Conference on Empirical Methods in Natural Language Processing and the 9th International Joint Conference on Natural Language Processing*, pages 2153–2162.

Jiuniu Wang, Wenjia Xu, Xingyu Fu, Guangluan Xu, and Yirong Wu. 2020. Astral: adversarial trained lstm-cnn for named entity recognition. *Knowledge-Based Systems*, 197:105842.

Zhengli Zhao, Dheeru Dua, and Sameer Singh. 2018. Generating natural adversarial examples. In *6th International Conference on Learning Representations*.

GreaseVision: Rewriting the Rules of the Interface

Siddhartha Datta
University of Oxford
siddhartha.datta@cs.ox.ac.uk

Konrad Kollnig
University of Oxford
konrad.kollnig@cs.ox.ac.uk

Nigel Shadbolt
University of Oxford
nigel.shadbolt@cs.ox.ac.uk

Abstract

Digital harms can manifest across any interface. Key problems in addressing these harms include the high individuality of harms and the fast-changing nature of digital systems. We put forth *GreaseVision*, a collaborative human-in-the-loop learning framework that enables end-users to analyze their screenomes to annotate harms as well as render overlay interventions. We evaluate HITL intervention development with a set of completed tasks in a cognitive walkthrough, and test scalability with one-shot element removal and fine-tuning hate speech classification models. The contribution of the framework and tool allow individual end-users to study their usage history and create personalized interventions. Our contribution also enables researchers to study the distribution of multi-modal harms and interventions at scale.

1 Introduction

The design of good user interfaces can be challenging. In a fast changing world, however, with sometimes highly individual needs, traditional one-fits-all software development faces difficulty in keeping up with the pace of change and the breadth of user requirements. At the same time, the digital world is rife with a range of harms, from dark patterns to hate speech to violence. This paper takes a step back to improve the user experience in the digital world. To achieve this, we put forward a new design philosophy for the development of software interfaces that serves its users: *GreaseVision*.

Contributions: Our work aims to contribute a novel interface modification framework, which we call *GreaseVision*. At a structural-level, our framework enables end-users to develop personalized interface modifications, either individually or collaboratively. This is supported by the use of screenome visualization, human-in-the-loop learning, and an overlay/hooks-enabled low-code development platform. Within the defined scopes, we enable the aggregation of distributionally-wide end-user digital harms (self-reflection for end-users, or analyzing the harms dataset of text, images and elements for researchers), to further enable the modification of user interfaces across a wide range of software systems, supported by the usage of visual overlays, autonomously developed by users, and enhanced by scalable machine learning techniques. We provide complete and reproducible implementation details to enable researchers to not only study harms and interventions, but other interface modification use cases as well.

Structure: We introduce the challenge of end-user interface modification in Sections 1 and 2 to curb digital harms. We share our proposed method – *GreaseVision* – in Section 3. We evaluate our method in Section 4, and share final thoughts and conclusions in Section 5.

2 Related Works & Problem

We summarize key related work here; for detailed related works, we refer the reader to Appendix: Section 6.2. Our motivating problem is the high individuality of digital harms across a distribution of users. The harms landscape is quickly changing with ever-changing digital systems, ranging from heavily-biased content (e.g. disinformation, hate speech), self-harm (e.g. eating disorders, self-cutting, suicide), cyber crime (e.g. cyberbullying, harassment, promotion of and recruitment for extreme causes (e.g. terrorist organizations), to demographic-specific exploitation (e.g. child-inappropriate content, social engineering attacks) (HM, 2019; Pater and Mynatt, 2017; Wang et al., 2017; Honary et al., 2020; Pater et al., 2019). Though interface modification frameworks exist, the distribution of the interface modifications (*interventions*) are constrained to the development efforts of an intervention developer, the availability of interventions are skewed towards desktop browsers (much sparser on mobile), and the efforts

Proceedings of the First Workshop on Dynamic Adversarial Data Collection, pages 7 - 22
July 14, 2022 ©2022 Association for Computational Linguistics

Figure 1: Walkthrough of using *GreaseVision*-modified interfaces.

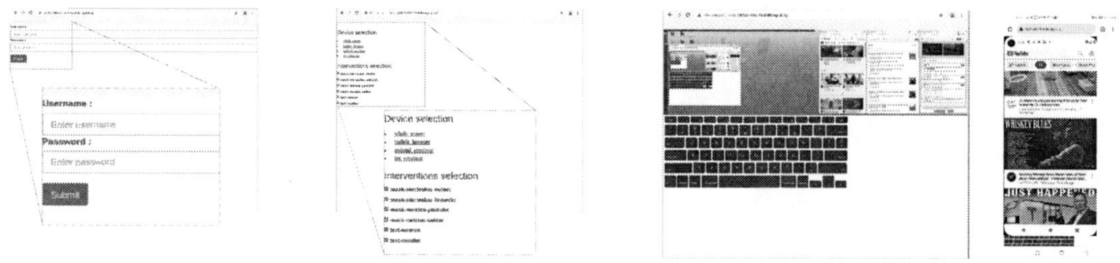

(a) *User authentication*: Secure gateway to the user's screenomes, personal devices, and intervention development suite.

(b) *Interface & interventions selection*: Listings of all registered devices/emulators on server, as well as interventions contributed by the users or community using the tool in Figure 3.

(c) *Interface access*: Accessing a Linux desktop from another (Linux) desktop browser.

(d) *Interface access*: Accessing an Android emulator from another Android host device.

are highly interface-specific (an app version update breaks code-based modification; videos cannot be perturbed in real-time). Moreover, low-code development platforms, that enable end-users to use visual tools to construct programs, are mostly available for software creation, but few options exist for software modification.

Due to the non-uniform distribution of users, the diverging distribution of harms would require a wide distribution of interventions. We hypothesize we can generate this matching distribution of interventions by enabling end-users to render personalized interventions *by themselves* (i.e. removing intervention developers from the ecosystem). To test this hypothesis, we attempt to bind the harms landscape to the interventions landscape by developing a collaborative human-in-the-loop system where end-users can inspect their browsing history and generate corresponding interventions.

We pursue a *visual overlay modifications* approach, extending on the work of ***GreaseTerminator*** (Datta et al., 2021). The framework renders overlay graphics over an underlay screen based on detected GUI elements, images or text (as opposed to implementing program code changes natively), hence changes the interface rather than the functionality of the software. To provide end-users with input for self-reflection (Cho et al., 2021; Lyngs et al., 2020a) and source data for generating interventions, users can be shown their **screenome** (Reeves et al., 2020, 2021), a record of a user's digital experiences represented as a sequence of screen images that they view and interact with over time To connect the input (screenome) and output (intervention), **Human-in-the-Loop (HITL)** learning can be used for users to annotate their screenomes for harmful text, images or GUI elements, and these annotations can be used to develop interventions. Wu et al. (2021) offers a detailed review of HITL. Specifically, the procedure to generate interventions using visual overlays will require the one-shot detection of masks (e.g. GUI elements) and few-shot learning and/or model fine-tuning of image/text classification models.

System requirements: Based on the problem and our collaborative HITL interface modification approach, we establish the following technical requirement (Requirement 1) and systemic requirement (Requirement 2) for our framework:

1. **(Req 1)** *A complete feedback loop between user input (train-time) and interface re-render (test-time).*
2. **(Req 2)** *Prospects for scalability across the distribution of interface modifications (with respect to both harms landscape and rendering landscape).*

3 *GreaseVision*

3.1 System Architecture: Binding the Harms Ecosystem to the Interventions Ecosystem

We define the *GreaseVision* architecture, with which end-users (system administrators) interact with, as follows (Figure 6(b)): (i) the user logs into the *GreaseVision* system to access amongst a set of personal emulators and interventions (the system admin has provisioned a set of emulated devices, hosted on a server through a set of virtual machines or docker containers for each emulator/interface,

Figure 2: Hooks adapted in *GreaseVision* to occlude distracting elements, censor hate speech, and obscure child-inappropriate content.

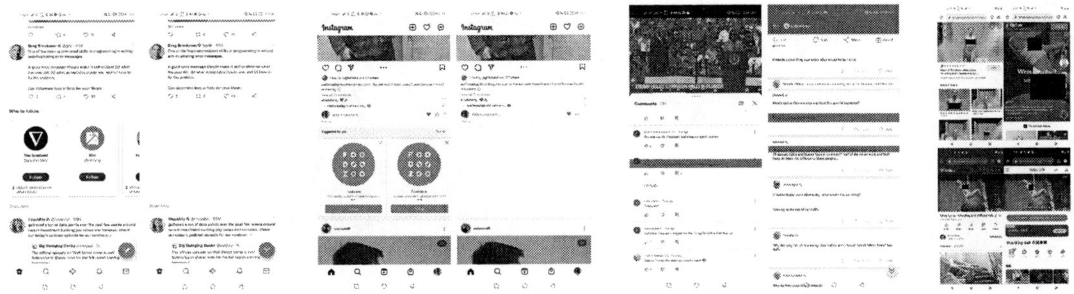

(a) Occlusion of recommended items on Twitter (before *left*, after *right*)

(b) Occlusion of recommended items on Instagram (before *left*, after *right*)

(c) Text censoring (YouTube *left*, Reddit *right*)

(d) Content moderation (Google Images, TikTok, YouTube, YouKu)

and handling streaming of the emulators, handling pre-requisites for the emulators, handling data migrations, etc); (ii) the user selects their desired interventions and continues browsing on their interfaces; (iii) after a time period, the user accesses their screenome and annotates interface elements, graphics, or text that they would like to generate interventions off of, which then re-populate the list of interventions available to members in a network.

In addition to the contributions stated in Section 1, *GreaseVision* is an improved visual overlay modification approach with respect to interface-agnosticity and ease of use. We discuss the specific aspects of *GreaseTerminator* we adopt in *GreaseVision* (hooks and overlays), and the technical improvements upon *GreaseTerminator* in Appendix: Section 6.1, specifically latency, device support, and interface-agnosticity.

In our current implementation, the user accesses a web application (compatible with both desktop and mobile browsers). With their login credentials, the database loads the corresponding mapping of the user's virtual machines/containers that are shown in the interface selection page. The central server carries information on accessing a set of emulated devices (devices loaded locally on the central server in our setup). Each emulator is rendered in docker containers or virtual machines where input commands can be redirected. The database also loads the corresponding mapping of available interventions (generated by the user, or by the network of users) in the interventions selection page. The database also loads the screenomes (images of all timestamped, browsed interfaces) in the screenome visualization page. Primary input commands for both desktop and mobile are encoded, including keystroke entry (hardware keyboard, on-screen keyboard), mouse/touch input (scrolling, swiping, pinching, etc); input is locked to the coordinates of the displayed screen image on the web app (to avoid stray/accidental input commands), and the coordinates correspond to each virtual machine/container's display coordinates. Screen images are captured at a configurable framerate (we set it to 60FPS), and the images are stored under a directory mapped to the user. Generated masks and fine-tuned models are stored under an interventions directory and their intervention/file access is also locked by mapped users. Interventions are applied sequentially upon a screen image to return a perturbed/new image, which then updates the screen image shown on the client web app.

3.2 Low-code Development: Binding Screenomes to Interface Modifications

We make use of the three hooks from *GreaseTerminator* (*text*, *mask*, and *model* hooks), and link it with the screenome visualization tool. While in *GreaseTerminator* the hooks ease the intervention development process for intervention developers with previous programming knowledge, we further generalize the intervention development process for intervention developers to the extent that even an end-user can craft their own interventions without developer support nor expert knowledge. *GreaseTerminator* enables intervention generation (via hooks) and interface re-rendering (via overlays). The added *GreaseVision* contribution of con-

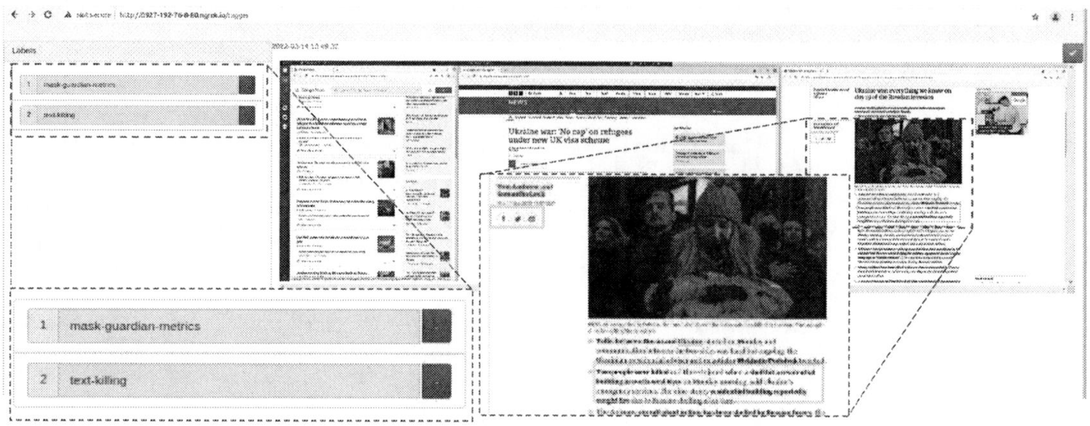

Figure 3: *Screenome visualization page*: The page offers the end-user the ability to traverse through the sequence of timestamped screen images which compose their screenome. They can use bounding boxes to highlight GUI elements, images or text. They can label these elements with specific encodings, such as mask- or text-.

necting these components with HITL learning and screenome visualization to replace developers is what exemplifies end-user autonomy and scalability in personalized interventions.

An intersecting data source that enables both end-user self-reflection (Cho et al., 2021; Lyngs et al., 2020a) and interface re-rendering via overlay (Datta et al., 2021) is the screenome. Specifically, we can orchestrate a loop that receives input from users and generates outputs for users. Through *GreaseVision*, end-users can browse through their own screen history, and beyond self-analysis, they can constructively build interface modifications to tackle specific needs. Extending on the interface rendering approach of overlays and hook-based intervention development, a generalizable design pattern for *GreaseTerminator*-based interventions is observed, where current few-shot/fine-tuning techniques can reasonably approach many digital harms, given appropriate extensions to the end-user development suite. In the current development suite (Figure 3), an end-user can inspect their screenomes across all *GreaseVision*-enabled interfaces (ranging from iOS, Android to desktops), and make use of image segment highlighting techniques to annotate interface patterns to detect (typically UI elements or image/text) and subsequently intervene against these interface patterns. Specifically, the interface images being stored and mapped to a user is shown in time-series sequence to the user. The user can go through the sequence of images to reflect on their browsing behavior. The current implementation focuses on one-shot detection of masks and fine-tuning of image and text classification models. When the user identifies a GUI element they do not wish to see across interfaces and apps, they highlight the region of the image, and annotate it as mask-<name-of-intervention>, and the mask hook will store a mask of intervention <name-of-intervention>, which will then populate a list of available interventions with this option, and the user can choose to activate it during a browsing session. When a user identifies text (images) that they do not wish to see of similar variations, they can highlight the text (image) region, and annotate it as text-<name-of-intervention> (image-<name-of-intervention>). The text hook will extract the text via OCR, and fine-tune a pretrained text classification model specifically for this type of text <name-of-intervention>. For images, the highlighted region will be cropped as input to fine-tune a pretrained image classification model. The corresponding text (image) occlusion intervention will censor similar text (images) during the user's browsing sessions if activated.

Extending on model few-shot training and fine-tuning, we can scale the accuracy of the models, not just through improvements to these training methods, but also by improving the data collection dynamics. More specifically, based on the spectrum of personalized and overlapping intervention needs for a distribution of users, we can leverage model-human and human-human collaboration to scale the generation of mask and model interventions. In the case of mask hooks, end-users who encounter certain harmful GUI elements (perhaps due to exposure to specific apps or features prior to other users) can tag and share the mask intervention with other users collaboratively.

To collaboratively fine-tune models, users tag text based on a general intervention/category label, that is used to group text together to form a mini-dataset to fine-tune the model. An example of this would be a network of users highlighting racist text they come across in their screenomes that made them uncomfortable during their browsing sessions, and tagging them as `text-racist`, which aggregates more sentences to fine-tune a text classification model responsible for detecting/classifying text as racist or not, and subsequently occluding the text for the network of users during their live browsing sessions. The current premise is that users in a network know a ground-truth label of the category of the specific text they wish to detect and occlude, and the crowd-sourced text of each of N categories will yield corresponding N fine-tuned models. Collaborative labelling scales the rate in which text of a specific category can be acquired, reducing the burden on a single user while also diversifying the fine-tune training set, while also proliferating the fine-tuned models across a network of users and not wasting effort re-training already fine-tuned models of other users (i.e. increasing scalability of crafting and usage of interventions).

4 Evaluation

We evaluate the usability of (*Req 1*) the HITL component (usability for a *single* user with respect to inputs/outputs; or "does our system help generate interventions?"), and (*Req 2*) the collaborative component (improvement to usability for a *single* user when *multiple* users are involved; or "does our system scale with user numbers?") with **cognitive walkthroughs** and **scalability tests** respectively.

4.1 Cognitive Walkthrough

Qualitatively, we perform a cognitive walkthrough (John and Packer, 1995; Rieman et al., 1995) of the user experience to simulate the cognitive process and explicit actions taken by an end-user during usage of *GreaseVision* to access interfaces and craft interventions. In our walkthrough, we as researchers presume the role of an end-user. We state the walkthrough **step** in **bold**, *data pertaining to the task* in *italics*, and descriptive evaluation in normal font. To evaluate the process of constructing an intervention using our proposed HITL system, we inspect *the completion of a set of required tasks* based on criteria from Parasuraman et al.'s (Parasuraman et al., 2000) 4 types of automation applications, which aim to measure the role of automation in the harms self-reflection and intervention self-development process. The four required tasks to be completed are:

1. *Information Acquisition*: Could a user collect new data points to be used in intervention crafting?
2. *Information Analysis*: Could a user analyze interface data to inform them of potential harms and interventions?
3. *Decision & Action Selection*: Could a user act upon the analyzed information about the harms they are exposed to, and develop interventions?
4. *Action Implementation*: Could a user deploy the intervention in future browsing sessions?

User logs in (Figure 1a): *The user enters their username and password. These credentials are stored in a database mapped to a specific (set of) virtual machine(s) that contain the interfaces the user registered for access.* This is a standard step for any secured or personalized system, where a user is informed they are accessing data and information that is tailored for their own usage.

User selects active interface and interventions (Figure 1b): *The user is shown a set of available interventions, be it contributed by themselves or other users in a network. They select their target interventions, and select an interface to access during this session.* Based on their own configurations (e.g. *GreaseVision* set up locally on their own computer, or specific virtual machines set up for the required interfaces), users can view the set of interfaces that they can access and use to facilitate their digital experiences. The interface is available 24/7, retains all their personal data and storage, is recording their screenome data for review, and accessible via a web browser from any other device/platform. They are less constrained by the hardware limitations of their personal device, and just need to ensure the server of the interfaces has sufficient compute resources to host the interface and run the interventions. The populated interventions are also important to the user, as it is a marketplace and ecosystem of personalized and shareable interventions. Users can populate interventions that they themselves can generate through the screenome visualization tool, or access interventions collaboratively trained and contributed by multiple members in their network. The interventions are also modu-

Figure 4: Removal of GUI elements (YouTube sharing metrics/buttons) across multiple target interfaces and operating systems.

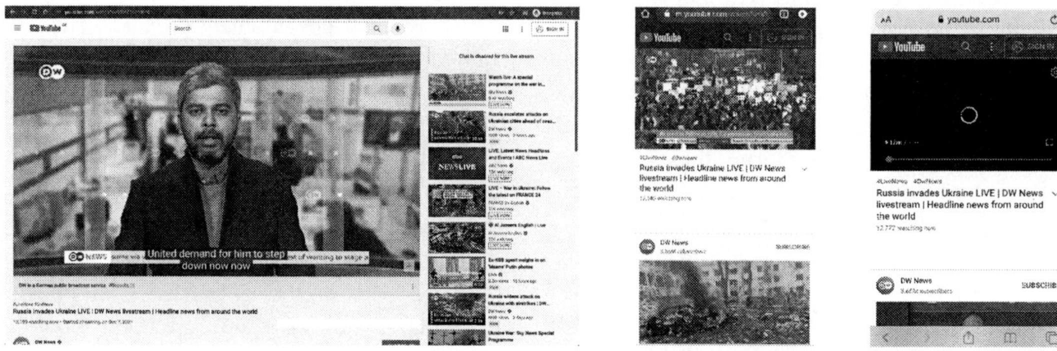

(a) Element removal on emulated desktop (MacOS)

(b) Element removal on emulated Android

(c) Element removal on emulated iOS

Mask	Min. masks	Android app	iOS app	Mobile browser	Desktop browser
Stories bar					
- Twitter	1	✓	✓	-	-
- Linkedin	1	✓	✓	-	-
- Instagram	1	✓	✓	-	-
Metrics/Sharing bar					
- Facebook	2	✓	✓	✓	✓
- Instagram	2	✓	✓	✓	✓
- Twitter	2	✓	✓	✓	✓
- YouTube	2	✓	✓	✓	✓
- TikTok	2	✓	✓	✓	✓
Recommended items					
- Twitter	2	✓	✓	✓	✓
- Facebook	2	✓	✓	✓	✓

Table 1: ✓ if element removal is successful, ✗ if element removal is unsuccessful, — if the element not available on an interface.

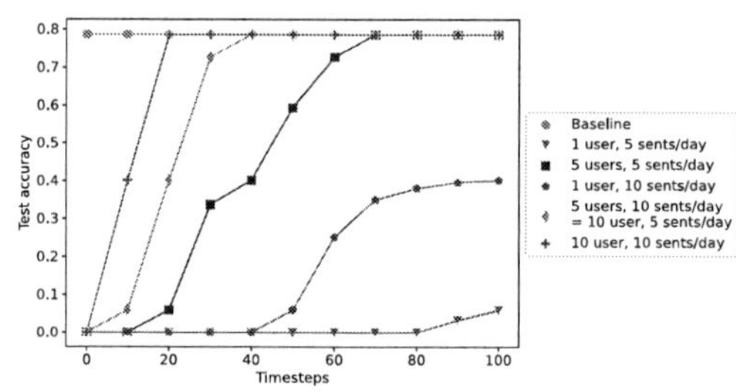

Figure 5: Convergence of few-shot/fine-tuned models on sub-groups of hate speech

lar enough that users are not restricted to a specific combination of interventions, and are applied sequentially onto the interface without mismatch in latency between the overlay and underlying interface. As the capabilities of generating interventions (e.g. more hooks) and rendering interfaces (e.g. interface augmentation) become extended, so do their ability to personalize their digital experience, and generate a distribution of digital experiences to match a similarly wide distribution of users. The autonomy to deploy interventions, with enhanced optionality through community-contributed interventions, before usage of an interface satisfies Task 4.

The user accesses the interface and browses (Figure 1c): *The user begins usage of the interface through the browser from their desired host device, be it mobile or desktop. They enter input to the system, which is streamed to the virtual machine(s), and interventions render overlay graphics to make any required interface modifications.* After the user has chosen their desired interventions, the user will enjoy an improved digital experience through the lack of exposure to certain digital elements, such as undesired text or GUI elements. The altered viewing experience satisfies both Task 1 and 4; not only is raw screen data being collected, but the screen is being altered by deployed interventions in the wild. The user cannot be harmed by what they previously chose not to see, and what they do see but no longer wish to see in the future, they can annotate to remove in future viewings in the screenome visualization tool. It is a cyclical loop where users can redesign and self-improve their browsing experiences through the use of unilateral or user-driven tools.

The user browses their screenome to generate interventions (Figure 3): *After a browsing period, the user may opt to browse and view their personal screenome. They enter the screenome visualization page to view recorded intervals of their browsing activity across all interfaces, and they can choose to annotate certain regions (image or text) to generate interventions to re-populate the interventions available.* The user is given autonomy in selecting and determining what aspects of the interface, be it the static app interface of dynamic content provisioned, that they no longer wish to see in the future. Enabling the user to view their screenome across all used digital interfaces (extending to mobile and desktop) to self-reflect and analyze browsing or content patterns satisfies Task 2. Though the screenome provides the user raw historical data, it may require additional processing (e.g. automated analysis, charts) to avoid information overload. Rather than waiting for a feedback loop for the app/platform developers or altruistic intervention developers to craft broad-spectrum interventions that may or may not fit their personal needs, the end-user can enjoy a personalized loop of crafting and deploying interventions, almost instantly for certain interventions such as element masks. The user can enter metadata pertaining to each highlighted harm, and not only contribute to their own experience improvement, but also contribute to the improvement of others who may not have encountered or annotated the harm yet. By developing interventions based on their analysis, not only for themselves but potentially for other users, they could successfully achieve Task 3. Though previously-stated as out of scope, to further support Task 3, other potential intervention modalities such as augmentation could also be contributed by a community of professional intervention developers/researchers (who redirect efforts from individual interventions towards enhancing low-code development tools).

The four tasks, used to determine whether a complete feedback loop between input collection/processing and interface rendering through HITL by a single user, could all be successfully completed, thus *GreaseVision* satisfies Requirement 1.

4.2 Scalability Testing

To evaluate the collaborative component, we measure the improvement to the user experience of a *single* user through the efforts of multiple users. We evaluate through scalability testing (Meerts and Graham, 2010), a type of load testing that measures a system's ability to scale with respect to the number of users. We simulate the usage of the system to evaluate the scalable generation of one-shot graphics (mask) detection, and scalable fine-tuning/few-shot training of (text) models. We do not replicate the scalability analysis on real users: the fine-tuning mechanism is still the same, and the main variable (in common) is the sentences highlighted (and their assigned labels and metadata, as well as the quality of the annotations), though error is expectedly higher in the real-world as the data may be sampled differently and of lower annotation quality. The primary utility of collaboration to an individual end-user is the scaled reduction of effort in intervention development. We evaluate this in terms of variety of individualized interventions (variations of masks), and the time saved in constructing a single robust intervention (time needed to construct an accurate model intervention).

Breadth of interface-agnostic masks (Table 1): We investigate the ease to annotate graphically-consistent GUI elements for few-shot detection. We sample elements to occlude that can exist across a variety of interfaces. We evaluate the occlusion of the *stories bar* (pre-dominantly only found on mobile devices, not desktop/browsers); some intervention tools exist on Android (Happening, 2021; MaaarZ, 2019; Kollnig et al., 2021; Datta et al., 2021) and iOS (Friendly, 2022), though the tools are app- (and version-) specific. We evaluate the occlusion of *like/share metrics*; there are mainly desktop browser intervention tools (Grosser, 2012, 2018, 2019; hidelikes, 2022), and one Android intervention tool (Datta et al., 2021). We evaluate the occlusion of *recommendations*; there are intervention tools that remove varying extents of the interface on browsers (such as the entire newsfeed) (West, 2012; Unhook, 2022). Existing implementations and interest in such interventions indicate some users have overlapping interests in tackling the removal or occlusion of such GUI elements, though the implementations may not exist across all interface platforms, and may not be robust to version changes. For each intervention, we evaluate on a range of target (emulated) interfaces. We aim for the real-time occlusion of the specific GUI element, and evaluate on native apps (for Android and iOS) and browsers (Android mobile browser, and Linux desktop browser).

For each of the GUI element cases, we make use of the screenome visualization tool to annotate and tag the minimum number of masks of the specific elements we wish to block across a set of apps. There tend to be small variations in the design of the element between browsers and mobile, hence we tend to require at least 1 mask from each device type; Android and iOS apps tend to have similar enough GUI elements that a single mask can be reused between them. We tabulate in Table 1 the successful generation and real-time occlusion of all evaluated and applicable GUI elements. We append screenshots of the removal of recommended items from the Twitter and Instagram apps on Android (Figure 2(a,b)). We append screenshots of the demetrification (occlusion of like/share buttons and metrics) of YouTube across desktop browsers (MacOS) and mobile browsers (Android, iOS) (Figure 4).

Convergence of few-shot/fine-tune trained text models (Figure 5): We investigate the accuracy gains from fine-tuning pretrained text models as a function of user numbers and annotated sentence contributions. Specifically, we evaluate the text censoring of hate speech, where the primary form of mitigation is still community standard guidelines and platform moderation, with a few end-user tooling available (Bodyguard, 2019; Datta et al., 2021). The premise of this empirical evaluation is that we have a group of simualated users N who each contribute N inputs (sentences) of a specific target class (hate speech, specifically against women) per timestep. With respect to a baseline, which is a pretrained model fine-tuned with all available sentences against women from a hate speech dataset, we wish to observe how the test accuracy of a model fine-tuned with $M \times N$ sentences varies over time. Our source of hate speech for evaluation is the *Dynamically Generated Hate Speech Dataset* (Vidgen et al., 2021), which contains sentences of `non-hate` and `hate` labels, and also classifies hate-labelled data by the target victim of the text (e.g. `women`, `muslim`, `jewish`, `black`, `disabled`). As we expect the M users to be labelling a specific niche of hate speech to censor, we specify the subset of hate speech of `women` (train set count: 1,652; test set count: 187). We fine-tune a publicly-available, pre-trained *RoBERTa* model (HuggingFace, 2022; Liu et al., 2019), which was trained on a large corpus of English data (*Wikipedia* (Wikimedia), *BookCorpus* (Zhu et al., 2015)). For each constant number of users M and constant sentence sampling rate N, at each timestep t, $M \times N \times t$ sentences are acquired of class `hate` against target `women`; there are a total of 1,652 train set sentences under these constraints (i.e. the max number of sentences that can be acquired before it hits the baseline accuracy), and to balance the class distribution, we retain all 15,184 train set `non-hate` sentences. We evaluate the test accuracy of the fine-tuned model on all 187 test set women-targeted hate speech. We also vary M and N to observe sensitivity of these parameters to the convergence towards baseline test accuracy.

The rate of convergence of a finetuned model is quicker when the number of users and contributed sentences per timestep both increase, approximately when we reach at least 1,000 sentences for the `women` hate speech category. The difference in convergence rates indicate that a collaborative approach to training can scale interventions development, as opposed to training text classification models from scratch and each user annotating text alone.

The empirical results for this section are stated in Table 1 and Figure 5. The data and evaluations from the scalability tests indicate that the ease of mask generation and model fine-tuning, further catalyzed by performance improvements from more users, enable the scalable generation of interventions and their associated harms, thus *GreaseVision* satisfies Requirement 2.

5 Conclusion

To enable end-user autonomy over interface design, and the generation and proliferation of a distribution of harms and interventions to analyze and reflect upon, we contribute the novel interface modification framework *GreaseVision*. End-users can reflect and annotate with their digital browsing experiences, and collaboratively craft interface interventions with our HITL and visual overlay mechanisms. With respect to Requirements 1 and 2, we find that our *GreaseVision* framework allows for scalable yet personalized end-user development of interventions against element, image and text-based digital harms. We hope *GreaseVision* will enable researchers and end-users to study harms and interventions, and other interface modification use cases.

References

Samira Abnar, Mostafa Dehghani, Behnam Neyshabur, and Hanie Sedghi. 2022. Exploring the limits of large scale pre-training. In *International Conference on Learning Representations*.

Yuvraj Agarwal and Malcolm Hall. 2013. ProtectMyPrivacy: Detecting and mitigating privacy leaks on iOS devices using crowdsourcing. In *Proceeding of the 11th Annual International Conference on Mobile Systems, Applications, and Services - MobiSys '13*, page 97, Taipei, Taiwan. ACM Press.

Armen Aghajanyan, Sonal Gupta, and Luke Zettlemoyer. 2021. Intrinsic dimensionality explains the effectiveness of language model fine-tuning. In *Proceedings of the 59th Annual Meeting of the Association for Computational Linguistics and the 11th International Joint Conference on Natural Language Processing (Volume 1: Long Papers)*, pages 7319–7328, Online. Association for Computational Linguistics.

Ionut Andone, Konrad Błaszkiewicz, Mark Eibes, Boris Trendafilov, Christian Montag, and Alexander Markowetz. 2016. Menthal: A framework for mobile data collection and analysis. In *Proceedings of the 2016 ACM International Joint Conference on Pervasive and Ubiquitous Computing: Adjunct*, UbiComp '16, page 624–629, New York, NY, USA. Association for Computing Machinery.

Michael Backes, Sebastian Gerling, Christian Hammer, Matteo Maffei, and Philipp von Styp-Rekowsky. 2014. AppGuard – Fine-Grained Policy Enforcement for Untrusted Android Applications. In Joaquin Garcia-Alfaro, Georgios Lioudakis, Nora Cuppens-Boulahia, Simon Foley, and William M. Fitzgerald, editors, *Data Privacy Management and Autonomous Spontaneous Security*, volume 8247 of *Lecture Notes in Computer Science*, pages 213–231. Springer Berlin Heidelberg, Berlin, Heidelberg.

Inc Bodyguard. 2019. Bodyguard.

Hyunsung Cho, DaEun Choi, Donghwi Kim, Wan Ju Kang, Eun Kyoung Choe, and Sung-Ju Lee. 2021. Reflect, not regret: Understanding regretful smartphone use with app feature-level analysis. *Proc. ACM Hum.-Comput. Interact.*, 5(CSCW2).

Siddhartha Datta. 2021. Learn2weight: Weights transfer defense against similar-domain adversarial attacks.

Siddhartha Datta, Konrad Kollnig, and Nigel Shadbolt. 2021. Mind-proofing your phone: Navigating the digital minefield with greaseterminator. *CoRR*, abs/2112.10699.

Siddhartha Datta and Nigel Shadbolt. 2022. Low-loss subspace compression for clean gains against multi-agent backdoor attacks. *arXiv preprint arXiv:2203.03692*.

Benjamin Davis and Hao Chen. 2013. RetroSkeleton: Retrofitting android apps. In *Proceeding of the 11th Annual International Conference on Mobile Systems, Applications, and Services - MobiSys '13*, page 181, Taipei, Taiwan. ACM Press.

Benjamin Davis, Ben S, Armen Khodaverdian, and Hao Chen. 2012. I-arm-droid: A rewriting framework for in-app reference monitors for android applications. In *In Proceedings of the Mobile Security Technologies 2012, MOST '12.*, pages 1–9, New York, NY, United States. IEEE.

William Enck, Peter Gilbert, Byung-Gon Chun, Landon P. Cox, Jaeyeon Jung, Patrick McDaniel, and Anmol N. Sheth. 2010. TaintDroid: An Information-flow Tracking System for Realtime Privacy Monitoring on Smartphones. In *Proceedings of the 9th USENIX Conference on Operating Systems Design and Implementation*, OSDI'10, pages 393–407, Berkeley, CA, United States. USENIX Association.

Chelsea Finn, Pieter Abbeel, and Sergey Levine. 2017. Model-agnostic meta-learning for fast adaptation of deep networks.

flxapps. 2021. Detoxdroid.

Jay Freeman. 2020. Cydia substrate.

App Studio Friendly. 2022. Friendly social browser.

Tomer Galanti, András György, and Marcus Hutter. 2022. On the role of neural collapse in transfer learning. In *International Conference on Learning Representations*.

Kovacs Geza. 2019. HabitLab: In-The-Wild Behavior Change Experiments at Scale. *Stanford Department of Computer Science*.

Vegard IT GmbH. 2021. Gray-switch.

Google. 2007. Tesseract.

Google. 2010a. Chrome web store.

Google. 2010b. recaptcha faq.

Google. 2021. Android accessibility suite.

Colin M. Gray, Yubo Kou, Bryan Battles, Joseph Hoggatt, and Austin L. Toombs. 2018. The dark (patterns) side of ux design. In *Proceedings of the 2018 CHI Conference on Human Factors in Computing Systems*, CHI '18, page 1–14, New York, NY, USA. Association for Computing Machinery.

Benjamin Grosser. 2012. Facebook demetricator.

Benjamin Grosser. 2018. Twitter demetricator.

Benjamin Grosser. 2019. Instagram demetricator.

Studios Happening. 2021. Swipe for facebook.

hidelikes. 2022. Hide likes.

Niklas Higi. 2020. apk-mitm.

Alexis Hiniker, Sungsoo (Ray) Hong, Tadayoshi Kohno, and Julie A. Kientz. 2016. Mytime: Designing and evaluating an intervention for smartphone non-use. In *Proceedings of the 2016 CHI Conference on Human Factors in Computing Systems*, CHI '16, page 4746–4757, New York, NY, USA. Association for Computing Machinery.

Government HM. 2019. Online Harms White Paper. *Government Report on Transparency Reporting*.

Mahsa Honary, Beth Bell, Sarah Clinch, Julio Vega, Leo Kroll, Aaron Sefi, and Roisin McNaney. 2020. Shaping the design of smartphone-based interventions for self-harm. In *Proceedings of the 2020 CHI Conference on Human Factors in Computing Systems*, CHI '20, page 1–14, New York, NY, USA. Association for Computing Machinery.

HuggingFace. 2022. roberta-base.

Andrey Ignatov. 2021. Ai-benchmark.

Jinseong Jeon, Kristopher K. Micinski, Jeffrey A. Vaughan, Ari Fogel, Nikhilesh Reddy, Jeffrey S. Foster, and Todd Millstein. 2012. Dr. Android and Mr. Hide: Fine-grained permissions in android applications. In *Proceedings of the Second ACM Workshop on Security and Privacy in Smartphones and Mobile Devices - SPSM '12*, page 3, Raleigh, North Carolina, USA. ACM Press.

Bonnie E. John and Hilary Packer. 1995. Learning and using the cognitive walkthrough method: A case study approach. In *Proceedings of the SIGCHI Conference on Human Factors in Computing Systems*, CHI '95, page 429–436, USA. ACM Press/Addison-Wesley Publishing Co.

Minsam Ko, Subin Yang, Joonwon Lee, Christian Heizmann, Jinyoung Jeong, Uichin Lee, Daehee Shin, Koji Yatani, Junehwa Song, and Kyong-Mee Chung. 2015. Nugu: A group-based intervention app for improving self-regulation of limiting smartphone use. In *Proceedings of the 18th ACM Conference on Computer Supported Cooperative Work & Social Computing*, CSCW '15, page 1235–1245, New York, NY, USA. Association for Computing Machinery.

Gregory Koch, Richard Zemel, and Ruslan Salakhutdinov. 2015. Siamese neural networks for one-shot image recognition.

Konrad Kollnig, Siddhartha Datta, and Max Van Kleek. 2021. I want my app that way: Reclaiming sovereignty over personal devices. In *Proceedings of the 2021 CHI Conference on Human Factors in Computing Systems Late-Breaking Works*, Yokohama, Japan. ACM Press.

AV Tech Labs. 2019. Auto logout.

Heyoung Lee, Heejune Ahn, Samwook Choi, and Wanbok Choi. 2014. The sams: Smartphone addiction management system and verification. *J. Med. Syst.*, 38(1):1–10.

Lawence Lessig. *Code 2.0*, 1 edition. Basic Books.

Yinhan Liu, Myle Ott, Naman Goyal, Jingfei Du, Mandar Joshi, Danqi Chen, Omer Levy, Mike Lewis, Luke Zettlemoyer, and Veselin Stoyanov. 2019. Roberta: A robustly optimized BERT pretraining approach. *CoRR*, abs/1907.11692.

Markus Löchtefeld, Matthias Böhmer, and Lyubomir Ganev. 2013. Appdetox: Helping users with mobile app addiction. In *Proceedings of the 12th International Conference on Mobile and Ubiquitous Multimedia*, MUM '13, New York, NY, USA. Association for Computing Machinery.

LuckyPatcher. 2020. Lucky patcher.

Yajing Luo, Peng Liang, Chong Wang, Mojtaba Shahin, and Jing Zhan. 2021. Characteristics and challenges of low-code development: The practitioners' perspective. *Proceedings of the 15th ACM / IEEE International Symposium on Empirical Software Engineering and Measurement (ESEM)*.

Ulrik Lyngs, Kai Lukoff, Petr Slovak, William Seymour, Helena Webb, Marina Jirotka, Jun Zhao, Max Van Kleek, and Nigel Shadbolt. 2020a. *'I Just Want to Hack Myself to Not Get Distracted': Evaluating Design Interventions for Self-Control on Facebook*, page 1–15. Association for Computing Machinery, New York, NY, USA.

Ulrik Lyngs, Kai Lukoff, Petr Slovak, William Seymour, Helena Webb, Marina Jirotka, Jun Zhao, Max Van Kleek, and Nigel Shadbolt. 2020b. 'I Just Want to Hack Myself to Not Get Distracted': Evaluating Design Interventions for Self-Control on Facebook. In *Proceedings of the 2020 CHI Conference on Human Factors in Computing Systems*, pages 1–15, Honolulu HI USA. ACM.

MaaarZ. 2019. Instaprefs.

Joris Meerts and Dorothy Graham. 2010. The history of software testing.

Meta. 2022. Content restrictions based on local law.

Behnam Neyshabur, Hanie Sedghi, and Chiyuan Zhang. 2020. What is being transferred in transfer learning? In *Advances in Neural Information Processing Systems*, volume 33, pages 512–523. Curran Associates, Inc.

Fabian Okeke, Michael Sobolev, Nicola Dell, and Deborah Estrin. 2018. Good vibrations: Can a digital nudge reduce digital overload? In *Proceedings of the 20th International Conference on Human-Computer Interaction with Mobile Devices and Services*, MobileHCI '18, New York, NY, USA. Association for Computing Machinery.

R. Parasuraman, T.B. Sheridan, and C.D. Wickens. 2000. A model for types and levels of human interaction with automation. *IEEE Transactions on Systems, Man, and Cybernetics - Part A: Systems and Humans*, 30(3):286–297.

Jessica Pater and Elizabeth Mynatt. 2017. Defining digital self-harm. In *Proceedings of the 2017 ACM Conference on Computer Supported Cooperative Work and Social Computing*, CSCW '17, page 1501–1513, New York, NY, USA. Association for Computing Machinery.

Jessica A. Pater, Brooke Farrington, Alycia Brown, Lauren E. Reining, Tammy Toscos, and Elizabeth D. Mynatt. 2019. Exploring indicators of digital self-harm with eating disorder patients: A case study. *Proc. ACM Hum.-Comput. Interact.*, 3(CSCW).

Aniruddh Raghu, Maithra Raghu, Samy Bengio, and Oriol Vinyals. 2020. Rapid learning or feature reuse? towards understanding the effectiveness of maml. In *International Conference on Learning Representations*.

Siegfried Rasthofer, Steven Arzt, Enrico Lovat, and Eric Bodden. 2014. DroidForce: Enforcing Complex, Data-centric, System-wide Policies in Android. In *2014 Ninth International Conference on Availability, Reliability and Security*, pages 40–49, Fribourg, Switzerland. IEEE.

Byron Reeves, Nilam Ram, Thomas N. Robinson, James J. Cummings, C. Lee Giles, Jennifer Pan, Agnese Chiatti, Mj Cho, Katie Roehrick, Xiao Yang, Anupriya Gagneja, Miriam Brinberg, Daniel Muise, Yingdan Lu, Mufan Luo, Andrew Fitzgerald, and Leo Yeykelis. 2021. Screenomics: A framework to capture and analyze personal life experiences and the ways that technology shapes them. *Human–Computer Interaction*, 36(2):150–201. PMID: 33867652.

Byron Reeves, Thomas Robinson, and Nilam Ram. 2020. Time for the human screenome project. *Nature*, 577(7790):314–317. Funding Information: The US National Institutes of Health (NIH) is Publisher Copyright: © 2020, Nature.

John Rieman, Marita Franzke, and David Redmiles. 1995. Usability evaluation with the cognitive walkthrough. In *Conference Companion on Human Factors in Computing Systems*, CHI '95, page 387–388, New York, NY, USA. Association for Computing Machinery.

rovo89. 2020. Xposed framework.

Christian Simon, Piotr Koniusz, Richard Nock, and Mehrtash Harandi. 2020. Adaptive subspaces for few-shot learning. In *2020 IEEE/CVF Conference on Computer Vision and Pattern Recognition (CVPR)*, pages 4135–4144.

Unhook. 2022. Unhook - remove youtube recommended videos.

Bertie Vidgen, Tristan Thrush, Zeerak Waseem, and Douwe Kiela. 2021. Learning from the worst: Dynamically generated datasets to improve online hate detection. In *Proceedings of the 59th Annual Meeting of the Association for Computational Linguistics and the 11th International Joint Conference on Natural Language Processing (Volume 1: Long Papers)*, pages 1667–1682, Online. Association for Computational Linguistics.

VrtualApp. 2016. Virtual xposed.

Eric Wallace, Pedro Rodriguez, Shi Feng, Ikuya Yamada, and Jordan Boyd-Graber. 2019. Trick me if you can: Human-in-the-loop generation of adversarial examples for question answering. *Transactions of the Association for Computational Linguistics*, 7:387–401.

Yilin Wang, Jiliang Tang, Jundong Li, Baoxin Li, Yali Wan, Clayton Mellina, Neil O'Hare, and Yi Chang. 2017. Understanding and discovering deliberate self-harm content in social media. In *Proceedings of the 26th International Conference on World Wide Web*, WWW '17, page 93–102, Republic and Canton of Geneva, CHE. International World Wide Web Conferences Steering Committee.

Jordan West. 2012. News feed eradicator for facebook.

Foundation Wikimedia. Wikimedia downloads.

Xingjiao Wu, Luwei Xiao, Yixuan Sun, Junhang Zhang, Tianlong Ma, and Liang He. 2021. A survey of human-in-the-loop for machine learning.

Rubin Xu, Hassen Saïdi, and Ross Anderson. 2012. Aurasium: Practical policy enforcement for android applications. In *21st USENIX Security Symposium (USENIX Security 12)*, pages 539–552, Bellevue, WA. USENIX Association.

Shanshan Zhang, Lihong He, Eduard Dragut, and Slobodan Vucetic. 2019. How to invest my time: Lessons from human-in-the-loop entity extraction. In *Proceedings of the 25th ACM SIGKDD International Conference on Knowledge Discovery & Data Mining*, KDD '19, page 2305–2313, New York, NY, USA. Association for Computing Machinery.

Xinyu Zhou, Cong Yao, He Wen, Yuzhi Wang, Shuchang Zhou, Weiran He, and Jiajun Liang. 2017. East: An efficient and accurate scene text detector.

Yukun Zhu, Ryan Kiros, Rich Zemel, Ruslan Salakhutdinov, Raquel Urtasun, Antonio Torralba, and Sanja Fidler. 2015. Aligning books and movies: Towards story-like visual explanations by watching movies and reading books. In *The IEEE International Conference on Computer Vision (ICCV)*.

6 Appendix

6.1 GreaseTerminator

In response to the continued widespread presence of interface-based harms in digital systems, Datta et al. (Datta et al., 2021) developed *GreaseTerminator*, a visual overlay modification method. This approach enables researchers to develop, deploy and study interventions against interface-based harms in apps. This is based on the observation that it used to be difficult in the past for researchers to study the efficacy of different intervention designs against harms within mobile apps (most previous approaches focused on desktop browsers). *GreaseTerminator* provides a set of 'hooks' that serve as templates for researchers to develop interventions, which are then deployed and tested with study participants. *GreaseTerminator* interventions usually come in the form of machine learning models that build on the provided hooks, automatically detect harms within the smartphone user interface at run-time, and choose appropriate interventions (e.g. a visual overlay to hide harmful content, or content warnings). The *GreaseTerminator* architecture is shown in Figure 6(a) in contrast to the *GreaseVision* architecture.

Technical improvements w.r.t. *GreaseTerminator* The improvements of *GreaseVision* with respect to *GreaseTerminator* are two-fold: (i) improvements to the framework enabling end-user development and harms mitigation (discussed in detail in Sections 4.2, 4.3, 5 and 6), and (ii) improvements to the technical architecture (which we discuss in this section). Our distinctive and non-trivial technical improvements to the *GreaseTerminator* architecture fall under namely latency, device support, and interface-agnosticity. *GreaseTerminator* requires the end-user device to be the host device, and overlays graphics on top. A downside of this is the non-uniformity of network latency between users (e.g. depending on the internet speed in their location) resulting in a potential mismatch in rendered overlays and underlying interface. With *GreaseVision*, we send a post-processed/re-rendered image once to the end-user device's browser (stream buffering) and do not need to send any screen image from the host user device to a server, thus there is no risk of overlay-underlay mismatch and we even reduce network latency by half. Images are relayed through an HTTPS connection, with a download/upload speed $\sim$ 250Mbps, and each image sent by the server amounting to $\sim$ 1Mb). The theoretical latency per one-way transmission should be $\frac{1 \times 1024 \times 8 \text{bits}}{250 \times 10^6 \text{bits/s}} = 0.033\text{ms}$. With each user at most requiring server usage of one NVIDIA GeForce RTX 2080, with reference to existing online benchmarks (Ignatov, 2021) the latency for 1 image (CNN) and text (LSTM) model would be 5.1ms and 4.8ms respectively. While the total theoretical latency for *GreaseTerminator* is $(2 \times 0.033 + 5)$, that of *GreaseVision* is $(0.033 + 5) = 5.03\text{ms}$. Another downside of *GreaseTerminator* is that it requires client-side software for each target platform. There would be pre-requisite OS requirements for the end-user device, where only versions of *GreaseTerminator* developed for each OS can be offered support (currently only for Android). *GreaseVision* streams screen images directly to a login-verified browser, allowing users to access desktop/mobile on any browser-supported device. Despite variations in the streaming architecture between *GreaseVision* and *GreaseTerminator*, the interface modification framework (hooks and overlays) are retained, hence interventions (even those developed by end-users) from *GreaseVision* are compatible in *GreaseTerminator*. In addition to improvements to the streaming architecture to fulfil interface-agnosticity, adapting the visual overlay modification framework into a collaborative HITL implementation further improves the ease-of-use for all stakeholders in the ecosystem. End-users do not need to root their devices, find intervention tools or even self-develop their own customized tools. We eliminate the need for researchers to craft interventions (as users self-develop autonomously) or develop their own custom experience sampling tools (as end-users/researchers can analyze digital experiences from stored screenomes). We also eliminate the need for intervention developers to learn a new technical framework or learn how to fine-tune models. Running emulators on docker containers and virtual machines on a (single) host server is feasible, and thus allows for the browser stream to be accessible cross-device without restriction, e.g. access iOS emulator on Android device, or macOS virtual machine on Windows device. Certain limitations are imposed on the current implementation, such as a lack of access to the device camera, audio, and haptics; however, these are not permanent issues, and engineered implementations exist where a virtual/emulated device can route and access the host device's input/output sources (VrtualApp, 2016).

Figure 6: Architecture of *GreaseTerminator* (left) and *GreaseVision* (right).

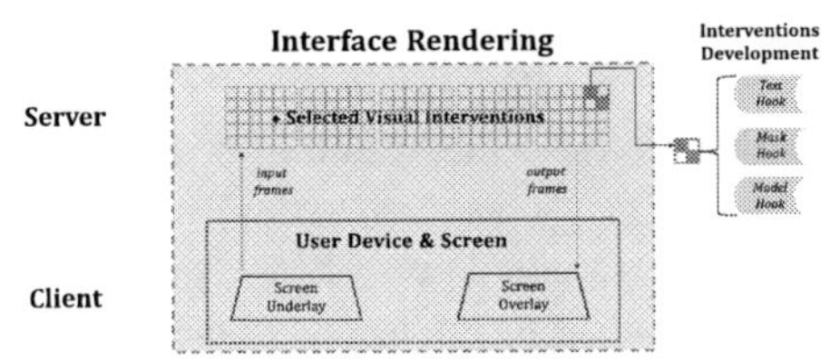

(a) The high-level architecture of *GreaseTerminator*. Details are explained in Section 2.3 and 4.2.

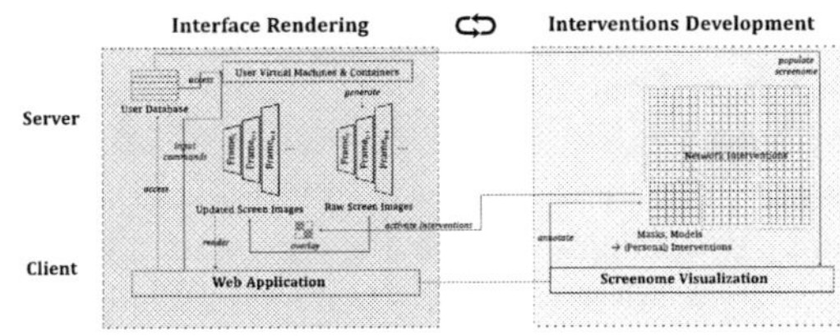

(b) The high-level architecture of *GreaseVision*, both as a summary of our technical infrastructure as well as one of the collaborative HITL interventions development approach.

Hooks The *text hook* enables modifying the text that is displayed on the user's device. It is implemented through character-level optical character recognition (OCR) that takes the screen image as an input and returns a set of characters and their corresponding coordinates. The EAST text detection (Zhou et al., 2017) model detects text in images and returns a set of regions with text, then uses Tesseract (Google, 2007) to extract characters within each region containing text. The *mask hook* matches the screen image against a target template of multiple images. It is implemented with *multi-scale multi-template* matching by resizing an image multiple times and sampling different subimages to compare against each instance of mask in a masks directory (where each mask is a cropped screenshot of an interface element). We retain the default majority-pixel inpainting method for mask hooks (inpainting with the most common colour value in a screen image or target masked region). As many mobile interfaces are standardized or uniform from a design perspective compared to images from the natural world, this may work in many instances. The mask hook could be connected to rendering functions such as highlighting the interface element with warning labels, or image inpainting (fill in the removed element pixels with newly generated pixels from the background), or adding content/information (from other apps) into the inpainted region. Developers can also tweak how the mask hook is applied, for example using the multi-scale multi-template matching algorithm with contourized images (shapes, colour-independent) or coloured images depending on whether the mask contains (dynamic) sub-elements, or using few-shot deep learning models if similar interface elements are non-uniform. A *model hook* loads any machine learning model to take any input and generate any output. This allows for model embedding (i.e. model weights and architectures) to inform further overlay rendering. We can connect models trained on specific tasks (e.g. person pose detection, emotion/sentiment analysis) to return output given the screen image (e.g. bounding box coordinates to filter), and this output can then be passed to a pre-defined rendering function (e.g. draw filtering box).

6.2 Related Works (extended)

6.2.1 Motivation: Pervasiveness and Individuality of Digital Harms

It is well-known that digital harms are widespread in our day-to-day technologies. Despite this, the academic literature around these harms is still developing, and it remains difficult to state exactly what the harms are that need to be addressed. Famously, Gray et al. (Gray et al., 2018) put forward a 5-class taxonomy to classify dark patterns within apps: *interface interference* (elements that manipulate the user interface to induce certain actions over other actions), *nagging* (elements that interrupt the user's current task with out-of-focus tasks) *forced action* (elements that introduce sub-tasks forcefully before permitting a user to complete their desired task), *obstruction* (elements that introduce subtasks with the intention of dissuading a user from performing an operation in the desired mode), and *sneaking* (elements that conceal or delay information relevant to the user in performing a task).

A challenge with such framework and taxonomies is to capture and understand the material impacts of harms on individuals. Harms tend to be highly individual and vary in terms of how they manifest within users of digital systems. The harms landscape is also quickly changing with ever-changing digital systems. Defining the spectrum of harms is still an open problem, the range varying from heavily-biased content (e.g. disinformation, hate speech), self-harm (e.g. eating disorders, self-cutting, suicide), cyber crime (e.g. cyber-bullying, harassment, promotion of and recruitment for extreme causes (e.g. terrorist organizations), to demographic-specific exploitation (e.g. child-inappropriate content, social engineering attacks) (HM, 2019; Pater and Mynatt, 2017; Wang et al., 2017; Honary et al., 2020; Pater et al., 2019), for which we recommend the aforementioned cited literature. The last line of defense against many digital harms is the user interface. This is why we are interested in interface-emergent harms in this paper, and how to support individuals in developing their own strategies to cope with and overcome such harms.

6.2.2 Developments in Interface Modification & Re-rendering

Digital harms have long been acknowledged as a general problem, and a range of technical interventions against digital harms are developed. Interventions, also similarly called modifications or patches, are changes to the software, which result in a change in (perceived) functionality and end-user usage. We review and categorize key *technical* intervention methods for interface modification by end-users, with cited examples specifically for digital harms mitigation. While there also exist non-technical interventions, in particular legal remedies, it is beyond this work to give a full account of these different interventions against harms; a useful framework for such an analysis is provided by Lawrence Lessig (Lessig) who characterised the different regulatory forces in the digital ecosystem.

Interface-code modifications (Kollnig et al., 2021; Higi, 2020; Jeon et al., 2012; Rasthofer et al., 2014; Davis and Chen, 2013; Backes et al., 2014; Xu et al., 2012; LuckyPatcher, 2020; Davis et al., 2012; Lyngs et al., 2020b; Freeman, 2020; rovo89, 2020; Agarwal and Hall, 2013; Enck et al., 2010; MaaarZ, 2019; VrtualApp, 2016) make changes to source code, either installation code (to modify software before installation), or run-time code (to modify software during usage). On desktop, this is done through *browser extensions* and has given rise to a large ecosystem of such extensions. Some of the most well-known interventions are ad blockers, and tools that improve productivity online (e.g. by removing the Facebook newsfeed (Lyngs et al., 2020b)). On mobile, a prominent example is *AppGuard* (Backes et al., 2014), a research project by Backes et al. that allowed users to improve the privacy properties of apps on their phone by making small, targeted modification to apps' source code. Another popular mobile solution in the community is the app *Lucky Patcher* (LuckyPatcher, 2020) that allows to get paid apps for free, by removing the code relating to payment functionality directly from the app code.

Some of these methods may require the highest level of privilege escalation to make modifications to the operating system and other programs/apps as a root user. On iOS, *Cydia Substrate* (Freeman, 2020) is the foundation for jailbreaking and further device modification. A similar system, called *Xposed Framework* (rovo89, 2020), exists for Android. To alleviate the risks and challenges afflicted with privilege escalation, *VirtualXposed* (VrtualApp, 2016) create a virtual environment on the user's Android device with simulated privilege escalation. Users can install apps into this virtual environment and apply tools of other modification approaches that may require root access. *ProtectMyPrivacy* (Agarwal and Hall, 2013) for iOS and *TaintDroid* (Enck et al., 2010) for Android both extend the functionality of the smartphone operating system with new functionality for the analysis of apps' privacy features. On desktops, code modifications tend not to be centred around a common framework, but are more commonplace in general due to the traditionally more permissive security model compared to mobile. Antivirus tools, copyright protections of games and the modding of UI components are all often implemented through interface-code modifications.

Interface-external modifications (Geza, 2019; Bodyguard, 2019; Lee et al., 2014; Ko et al., 2015; Andone et al., 2016; Hiniker et al., 2016; Löchtefeld et al., 2013; Labs, 2019; Okeke et al., 2018) are the arguably most common way to change default interface behaviour. An end-user would install a program so as to affect other programs/apps. No change to the operating system or the targeted programs/apps is made, so an uninstall of the program providing the modification would revert the device to the original state. This approach is widely used to track duration of device usage, send notifications to the user during usage (e.g. timers, warnings), block certain actions on the user device, and other aspects. The *HabitLab* (Geza, 2019) is a prominent example developed by Kovacs et al. at Stanford.

This modification framework is open-source and maintained by a community of developers, and provides interventions for both desktop and mobile.

Visual overlay modifications render graphics on an overlay layer over any active interface instance, including browsers, apps/programs, videos, or any other interface in the operating system. The modifications are visual, and do not change the functionality of the target interface. It may render sub-interfaces, labels, or other graphics on top of the foreground app. Prominent examples are *DetoxDroid* (flxapps, 2021), *Gray-Switch* (GmbH, 2021), *Google Accessibility Suite* (Google, 2021), and *GreaseTerminator* (Datta et al., 2021).

We would like to establish early on that we pursue a *visual overlay modifications* approach. Interventions should be rendered in the form of overlay graphics based on detected elements, rather than implementing program code changes natively, hence focused on changing the interface rather than the functionality of the software. Interventions should be generalizable; they are not solely website- or app-oriented, but *interface-oriented*. Interventions do not target specific apps, but general interface elements and patterns that could appear across different interface environments. To support the systemic requirements in Section 2.4, we require an interface modification approach that is (i) interface-agnostic and (ii) easy-to-use. To this extent, we build upon the work of ***GreaseTerminator*** (Datta et al., 2021), a framework optimized for these two requirements.

In response to the continued widespread presence of interface-based harms in digital systems, Datta et al. (Datta et al., 2021) developed *GreaseTerminator*, a visual overlay modification method. This approach enables researchers to develop, deploy and study interventions against interface-based harms in apps. This is based on the observation that it used to be difficult in the past for researchers to study the efficacy of different intervention designs against harms within mobile apps (most previous approaches focused on desktop browsers). *GreaseTerminator* provides a set of 'hooks' that serve as templates for researchers to develop interventions, which are then deployed and tested with study participants. *GreaseTerminator* interventions usually come in the form of machine learning models that build on the provided hooks, automatically detect harms within the smartphone user interface at run-time, and choose appropriate interventions (e.g. a visual overlay to hide harmful content, or content warnings). A visualisation of the *GreaseTerminator* approach is shown in Figure 6(a).

6.2.3 Opportunities for Low-code Development in Interface Modification

Low-code development platforms have been defined, according to practitioners, to be (i) low-code (negligible programming skill required to reach endgoal, potentially drag-and-drop), (ii) visual programming (a visual approach to development, mostly reliant on a GUI, and "what-you-see-is-what-you-get"), and (iii) automated (unattended operations exist to minimize human involvement) (Luo et al., 2021). Low-code development platforms exist for varying stages of software creation, from frontend (e.g. App maker, Bubble.io, Webflow), to workflow (Airtable, Amazon Honeycode, Google Tables, UiPath, Zapier), to backend (e.g. Firevase, WordPress, flutterflow); none exist for software modification of existing applications across interfaces. According to a review of StackOverflow and Reddit posts analysed by Luo et al. (Luo et al., 2021), low-code development platforms are cited by practitioners to be tools that enable faster development, lower the barrier to usage by non-technical people, improves IT governance compared to traditional programming, and even suits team development; one of the main limitations cited is that the complexity of the software created is constrained by the options offered by the platform.

User studies have shown that users can self-identify malevolent harms and habits upon self-reflection and develop desires to intervene against them (Cho et al., 2021; Lyngs et al., 2020a). Not only do end-users have a desire or interest in self-reflection, but there is indication that end-users have a willingness to act. Statistics for content violation reporting from Meta show that in the Jan-Jun 2021 period, $\sim$ 42,200 and $\sim$ 5,300 in-app content violations were reported on Facebook and Instagram respectively (Meta, 2022) (in this report, the numbers are specific to violations in local law, so the actual number with respect to community standard violatons would be much higher; the numbers also include reporting by governments/courts and non-government entities in addition to members of the public). Despite a willingness to act, there are limited digital visualization or reflection tools that enable flexible intervention development by end-users. There are visualization or reflection tools on browser and mobile that allow for reflection (e.g. device use time (Andone et al., 2016)), and there are separate and disconnected tools for intervention (Section 2.2), but there are limited offerings of flexible intervention development by end-users, where end-users can observe and analyze their problems while generating corresponding fixes, which thus prematurely ends the loop for action upon regret/reflection. There is a disconnect between the harms analysis ecosystem and interventions ecosystem. A barrier to binding these two ecosystems is the existence of low-code development platforms for end-users. While such tooling may exist for specific use cases on specific interfaces (e.g. web/app/game development) for mostly creationary purposes, there are limited options available for modification purposes of existing software, the closest alternative being extension ecosystems (Kollnig et al., 2021; Google, 2010a). Low-code development platforms are in essence "developer-less", removing developers from the software modification pipeline by reducing the barrier to modification through the use of GUI-based features and negligible coding, such that end-users can self-develop without expert knowledge.

Human-in-the-Loop (HITL) learning is the procedure of integrating human knowledge and experience in the augmentation of machine learning models. It is commonly used to generate new data from humans or annotate existing data by humans. Wallace et al. (Wallace et al., 2019) constructed a HITL system of an interactive interface where a human talks with a machine to generate more Q&A language and train/fine-tune Q&A models. Zhang et al. (Zhang et al., 2019) proposed a HITL system for humans to provide data for entity extraction, including requiring humans to formulate regular expressions and highlight text documents, and annotate and label data. For an extended literature review, we refer the reader to Wu et al. (Wu et al., 2021). Beyond lab settings, HITL has proven itself in wide deployment, where a wide distribution of users have indicated a willingness and ability to perform tasks on a HITL annotation tool, *reCAPTCHA*, to access utility and services. In 2010, Google reported over 100 million reCAPTCHA instances are displayed every day (Google, 2010b) to annotate different types of data, such as deciphering text for OCR of books or street signs, or labelling objects in images such as traffic lights or vehicles.

While HITL formulates the structure for human-AI collaborative model development, **model fine-tuning** and **few-shot learning** formulate the algorithmic methods of adapting models to changing inputs, environments, and contexts. Both adaptation approaches require the model to update its parameters with respect to the new input distribution. For model fine-tuning, the developer re-trains a pre-trained model on a new dataset. This is in contrast to training a model from a random initialization. Model fine-tuning techniques for pre-trained foundation models, that already contain many of the pre-requisite subnetworks required for feature reuse and warm-started training on a smaller target dataset, have indicated robustness on downstream tasks (Galanti et al., 2022; Abnar et al., 2022; Neyshabur et al., 2020). If there is an extremely large number of input distributions and few samples per distribution (small datasets), few-shot learning is an approach where the developer has separately trained a meta-model that learns how to change model parameters with respect to only a few samples. Few-shot learning has demonstrated successful test-time adaptation in updating model parameters with respect to limited test-time samples in both image and text domains (Raghu et al., 2020; Koch et al., 2015; Finn et al., 2017; Datta, 2021). Some overlapping techniques even exist between few-shot learning and fine-tuning, such as constructing subspaces and optimizing with respect to intrinsic dimensions (Aghajanyan et al., 2021; Datta and Shadbolt, 2022; Simon et al., 2020).

The raw data for harms and required interface changes reside in the history of interactions between the user and the interface. In the Screenome project (Reeves et al., 2020, 2021), the investigators proposed the study and analysis of the moment-by-moment changes on a person's screen, by capturing screenshots automatically and unobtrusively every $t = 5$ seconds while a device is on. This record of a user's digital experiences represented as a sequence of screens that they view and interact with over time is denoted as a user's **screenome**. Though not mobilized widely amongst users for their self-reflection or personalized analysis, integrating screenomes into an interface modification framework can play the dual roles of visualizing raw (harms) data to users while manifesting as parseable input for visual overlay modification frameworks.

Posthoc Verification and the Fallibility of the Ground Truth

Yifan Ding, Nicholas Botzer, Tim Weninger
Department of Computer Science & Engineering
University of Notre Dame
Notre Dame, IN, USA
{yding4,nbotzer,tweninge}@nd.edu

Abstract

Classifiers commonly make use of pre-annotated datasets, wherein a model is evaluated by pre-defined metrics on a held-out test set typically made of human-annotated labels. Metrics used in these evaluations are tied to the availability of well-defined ground truth labels, and these metrics typically do not allow for inexact matches. These noisy ground truth labels and strict evaluation metrics may compromise the validity and realism of evaluation results. In the present work, we conduct a systematic label verification experiment on the entity linking (EL) task. Specifically, we ask annotators to verify the correctness of annotations after the fact (*i.e.*, posthoc). Compared to pre-annotation evaluation, state-of-the-art EL models performed extremely well according to the posthoc evaluation methodology. Surprisingly, we find predictions from EL models had a similar or higher verification rate than the ground truth. We conclude with a discussion on these findings and recommendations for future evaluations. The source code, raw results, and evaluation scripts are publicly available via the MIT license at `https://github.com/yifding/e2e_EL_evaluate`

The general machine learning pipeline starts with a dataset (a collection of documents, images, medical records, etc.). When labels are not inherent to the data, they must be annotated – usually by humans. A label error occurs when an annotator provides a label that is "incorrect." But this raises an interesting question: who gets to decide that some annotation is incorrect?

One solution is to ask k annotators and combine their labels somehow (*e.g.*, majority vote, probability distribution). Subjectivity comes into play here. Given identical instructions and identical items, some annotators may focus on different attributes of the item or have a different interpretation of the labeling criteria. Understanding and modelling label uncertainty remains a compelling challenge in evaluating machine learning systems (Sommerauer, Fokkens, and Vossen, 2020; Resnick et al., 2021).

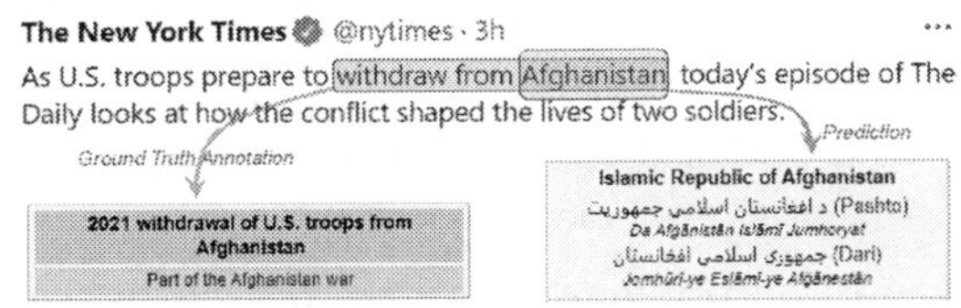

Figure 1: Example Entity Linking task where the pre-annotated ground truth mention and link is different from the predicted label. Standard evaluation regimes count this as a completely incorrect prediction despite being a reasonable label.

Tasks that require free-form, soft, or multi-class annotations present another dimension to this challenge. For example, natural language processing tasks like named entity recognition (NER) and entity linking (EL) rely heavily on datasets comprised of free-form human annotations. These tasks are typically evaluated against a held out portion of the already-annotated dataset. A problem arises when NER and EL tasks produce labels that are not easily verified as "close enough" to the correct groundtruth (Ribeiro et al., 2020). Instead, like the example in Fig. 1, most NER and EL evaluation metrics require exact matches against free-form annotations (Sevgili et al., 2020; Goel et al., 2021). This strict evaluation methodology may unreasonably count labels that are "close enough" as incorrect and is known to dramatically change performance metrics (Gashteovski et al., 2020).

Producing a *verifiable* answer is not the same as producing the *correct* answer. This distinction is critical. Asking a machine learning system to independently provide the same label as an annotator is a wildly different task than asking an annotator to verify the output of a predictor (*posthoc verification*). Unfortunately the prevailing test and evaluation regime requires predictors to exactly match noisy, free-form, and subjective human annotations. This paradigm represents a mismatch

Proceedings of the First Workshop on Dynamic Adversarial Data Collection, pages 23 - 29
July 14, 2022

Table 1: Statistics of the entity linking datasets and annotations.

	Datasets	Docs	Annotations			Tasks			Verified Annotations		
			GT	E2E	REL	GT	E2E	REL	GT	E2E	REL
AIDA	**AIDA-train**	946	18541*	18301	21204	2801	2802	2913	18511	18274	21172
	AIDA-A	216	4791	4758	5443	713	715	725	4787	4754	5439
	AIDA-B	231	4485	4375	5086	636	646	654	4480	4370	5079
WNED	**ACE2004**	57*	257	1355	1675	114	318	334	256	1352	1672
	AQUAINT	50	727	810	925	175	170	179	727	810	925
	CLUEWEB	320	11154	12273	23114	3526	3678	4944	11139	12247	23056
	MSNBC	20	656	629	756	164	163	171	656	629	756
	WIKIPEDIA	345*	6793*	8141	11184	1348	1578	1638	6786	8136	11177

* indicate results different from related work because they remove out-of-dictionary annotations.

that, if left unaddressed, threatens to undermine future progress in machine learning.

Main Contributions. We show that the distinction between pre-annotated and posthoc-annotated labels is substantial and the distinction presents consequences for how we determine the state-of-the-art in machine learning systems.

We conducted systematic experiments using posthoc analysis on a large case study of eight popular entity linking datasets with two state-of-the-art entity linking models, and report some surprising findings: First, state-of-the-art EL models generally predicted labels with *higher* verification rate than the ground truth labels. Second, there was substantial disagreement among annotators as to what constitutes a label that is "good enough" to be verified. Third, a large proportion (between 10%-70% depending on the dataset) of verified entities were missing from the ground truth dataset.

The Setting: Entity Linking

The goal of EL is to identify words or phrases that represent real-world entities and match each identified phrase to a listing in some knowledge base. Like most classification systems, EL models are typically trained and tested on large pre-annotated benchmark datasets. Table 1 describes eight such benchmark datasets that are widely used throughout the EL and broader NLP communities.

EL Models. In order to better understand the effect of pre-annotated benchmarks on machine learning systems, it is necessary to test a handful of state-of-the-art EL systems. Specifically, we chose: (1) The end-to-end (E2E) entity linking model, which generates and selects span candidates with associated entity labels. The E2E model is a word-level model that utilizes word and entity embeddings to compute span-level contextual

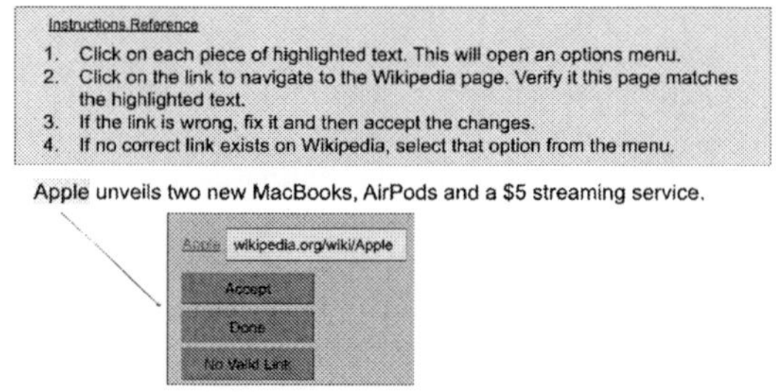

Figure 2: Web system used to collect posthoc annotations from workers.

scores. Word and entity embeddings are trained on Wikipedia, and the final model is trained and validated using AIDA-train and AIDA-A respectively (Kolitsas, Ganea, and Hofmann, 2018). (2) The Radboud Entity Linker (REL), which combines the Flair (Akbik, Blythe, and Vollgraf, 2018) NER system with the mulrel-nel (Le and Titov, 2018) entity disambiguation system to create a holistic EL pipeline (van Hulst et al., 2020). In addition, our methodology permits the evaluation of the GT as if it were a competing model. The relative performance of E2E and REL can then compared with the GT to better understand the performance of the posthoc annotations.

Data collection. We have previously argued that these evaluation metrics may not faithfully simulate *in vivo* performance because (1) the ground truth annotations are noisy and subjective, and (2) exact matching is too strict. We test this argument by collecting posthoc verifications of the three models, including the pre-annotated GT, over the datasets.

We created a simple verification system, illustrated in Fig. 2, and used Amazon Mechanical Turk to solicit workers. For each document and model, we asked a single worker to verify all present entity annotations (*i.e.*, an entity mention and its linked entity). Annotators can then choose to (1) Verify

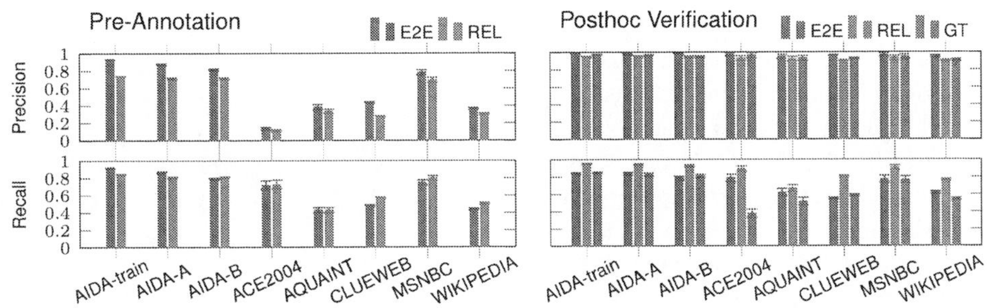

Figure 3: Precision and recall results from pre-annotation evaluation (Left) compared with the posthoc verification evaluation (Right). Error bars represent 95% confidence intervals on bootstrapped samples of the data. Posthoc verification returns substantially higher scores than the pre-annotation evaluation.

the annotation (2) Modify the annotation, or (3) Remove the annotation.

- **Verify**: The annotator determines that the current annotation (both mention and Wikipedia link) is appropriate.
- **Modify**: The annotator determines that the Wikipedia link is incorrect. In this case, they are asked to search and select a more appropriate Wikipedia link, use it to replace the existing link, and then accept the new annotation.
- **Remove**: The annotator determines that the current mention (highlighted text) is not a linkable entity. In this case, they remove the link from the mention.

We made a deliberate decision to not permit new annotation of missing entity mentions. That is, if the model did not label an entity, then there is no opportunity for the worker to add a new label. This design decision kept the worker focused on the verification task, but possibly limits the coverage of the verified dataset. We provide further comments on this decision in the Results section.

Each annotator is assigned to 20 tasks including one control task with three control annotations. We only accept and collect annotations from workers that passed the control task.

We paid each worker 3 USD for each HIT. We estimate a average hourly rate of about 9 USD; and paid a total of 6,520 USD. From these, we received 167,432 annotations. The breakdown of tasks, annotations shown to workers, and verified annotations are listed in Table 1 for each dataset and model.

Prior to launch, this experiment was reviewed and approved by an impaneled ethics review board at the University of Notre Dame. The source code, raw results, and evaluation scripts are publicly available via the MIT license at `https://github.com/yifding/e2e_EL_evaluate`

Posthoc Verification Methodology

The Pre-Annotation Evaluation Regime. First, we re-tested the E2E and REL models and evaluated their micro precision and recall under the typical pre-annotation evaluation regime. These results are illustrated in Fig 3 and are nearly identical to those reported by related works (Kolitsas, Ganea, and Hofmann, 2018; van Hulst et al., 2020).

Posthoc Verification Evaluation

Our next task is to define appropriate evaluation metrics that can be used to compare the results of the posthoc verification experiment with results from the pre-annotation evaluation regime.

Verification Rate. For each combination of dataset and model providing annotations, we compute the verification rate as the percentage of annotations that were verified. Formally, let $d \in$ datasets; $m \in$ models; and $V_{m,d}$ be the set of verified annotations in a pairing of d and m Likewise, let $N_{d,m}$ be the pre-annotations of model m on dataset d. We therefore define the verification rate of a dataset-model pair as $r_{m,d} = |V_{m,d}|/|N_{d,m}|$. Higher verification rates indicate that the dataset contains annotations and/or the model is more capable of providing labels that pass human inspection.

Verification Union. It is important to note that each model and document was evaluated by only a single worker. However, we were careful to assign each worker annotations randomly drawn from model/document combinations. This randomization largely eliminates biases in favor or against any model or dataset. Furthermore, this methodology provides for repetitions when annotations match exactly across models – which is what models are optimized for in the first place! In this scenario the union of all non-exact, non-overlapping annotations provides a superset of annotations similar to how pooling is used in information retrieval evaluation to create a robust result set (Zobel, 1998). Formally, we define the verification union of a dataset d as $V_d = \bigcup_m V_{m,d}$.

Posthoc Precision and Recall. The precision metric is defined as the ratio of true predictions to all predictions. If we recast the concept of true predictions to be the set of verified annotations $V_{m,d}$, then it is natural to further consider $N_{d,m}$ to be the set of all predictions for some dataset and model pair, especially considering our data collection methodology restricts $V_{m,d} \subseteq N_{d,m}$. Thus the posthoc precision of a model-data pairing is simply the verification rate $r_{m,d}$.

The recall metric is defined as the ratio of true predictions to all true labels. If we keep the recasting of true positives as verified annotations $V_{m,d}$, then all that remains a definition of true labels. Like in most evaluation regimes the set of all true labels is estimated by the available labels in the dataset. Here, we do the same and estimate the set of true labels as the union of a dataset's verified annotations V_d. Thus posthoc recall of a model-data pairing is $|V_{m,d}|/|V_d|$.

Posthoc Verification Results

Using the evaluation tools introduced in the previous section, we begin to answer interesting research questions. First, do the differences between evaluation regimes, *i.e.*, pre-annotation versus posthoc verification, have any affect on our perception of model performance.

To shed some light on this question, we compared the precision and recall metrics calculated using the pre-annotation evaluation regime against the precision and recall metrics calculated using the posthoc verification regime. The left quadplot in Fig. 3 compares model performance under the different evaluation regimes. Error bars represent the empirical 95% confidence internals drawn from 1000 bootstrap samples of the data. We make two major conclusions from this comparison:

Pre-annotation performance is lower than Posthoc verification. The differences between the scores of the pre-annotation compared to posthoc verification are striking. Posthoc annotation shows very good precision scores across all datasets. Although the models may not exactly predict the pre-annotated label, high posthoc precision indicates that their results appear to be "close-enough" to obtain human verification.

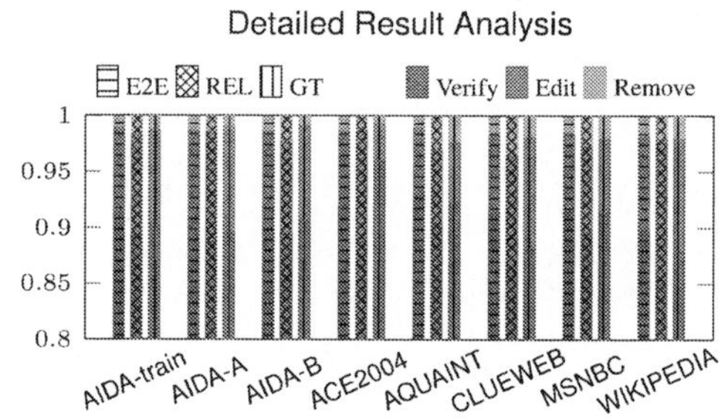

Figure 4: Detailed error analysis of verification rates in Fig. 3(top right). The E2E model consistently outperforms the ground truth (GT).

Conclusion: the widely-used exact matching evaluation regime is too strict. Despite its intention, the pre-annotation evaluation regime does not appear to faithfully simulate a human use case.

Machine Learning models outperform the Ground Truth. The posthoc verification methodology permits the GT annotations to be treated like any other model, and are therefore included in Fig. 3 (right plot). These results were unexpected and surprising. We found that labels produced by the EL models oftentimes had a higher verification rate than the pre-annotated ground truth. The recall metric also showed that the EL models were also able to identify more verified labels than GT.

Conclusion: Higher precision performance of the EL models indicates that human annotators make more unverifiable annotations than the EL models. Higher recall performance of the EL models also indicates that the EL models find a greater coverage of possible entities. The recall results are less surprising because human annotators may be unmotivated or inattentive during free-form annotation – qualities that tend to not affect EL models.

Error Analysis of the Ground Truth

For each linked entity, the posthoc verification methodology permitted one of three outcomes: verification, modification, or removal. The plot in Fig. 4 shows the percentage of each outcome for each model and dataset pair; it is essentially a zoomed-in, more-detailed illustration of the Posthoc Verification Precision result panel from Fig. 3, but with colors representing outcomes and patterns representing models. Edits indicate that the named entity recognition (*i.e.*, mention detection) portion of the EL model was able to identify an entity, but the entity was not linked to a verifiable entity. The available dataset has an enumeration of corrected linkages, but we do not consider them further in the present work. Removal indicates an error with the mention detection. From these results we find that, when a entity mention is detected it is usually a good detection; the majority of the error comes from the linking subtask.

A similar error analysis of missing entities is not permitted from the data collection methodology because we only ask workers to verify pre-annotated or predicted entities, not add missing entities. Because all detected mentions are provided with some entity link, we can safely assume that missing entities is mostly (perhaps wholly) due to errors in the mention detection portion of EL models.

Discussion

The primary goal of the present work is to compare pre-annotation labels contributed by human workers against verified annotations of the same data. Using entity linking as an example task, we ultimately found that these two methodologies returned vastly different performance results. From this observation we can draw several important conclusions. First, EL models have a much higher precision than related work reports. This difference is because the standard evaluation methodology used in EL, and throughout ML generally, do not account for soft matches or the semantics of what constitutes a label that is "close enough". Our second conclusion is that EL models, and perhaps ML models generally, sometimes perform better than ground truth annotators – at least, that is, according to other ground truth annotators.

Acknowledgments

This research is sponsored in part by the Defense Advanced Research Projects Agency (DAPRA) under contract numbers HR00111990114 and HR001121C0168. The views and conclusions contained herein are those of the authors and should not be interpreted as necessarily representing the official policies, either expressed or implied, of DARPA or the U.S. Government. The U.S. Government is authorized to reproduce and distribute reprints for governmental purposes notwithstanding any copyright annotation therein.

References

Aharoni, R.; and Goldberg, Y. 2018. Split and Rephrase: Better Evaluation and Stronger Baselines. In ACL, 719–724.

Akbik, A.; Blythe, D.; and Vollgraf, R. 2018. Contextual string embeddings for sequence labeling. In COLING, 1638–1649.

Belinkov, Y.; and Bisk, Y. 2018. Synthetic and Natural Noise Both Break Neural Machine Translation. In ICLR.

Botzer, N.; Ding, Y.; and Weninger, T. 2021. Reddit entity linking dataset. Information Processing & Management, 58(3): 102479.

Bowman, S. R.; Angeli, G.; Potts, C.; and Manning, C. D. 2015. A large annotated corpus for learning natural language inference. arXiv preprint arXiv:1508.05326.

Bowman, S. R.; and Dahl, G. E. 2021. What Will it Take to Fix Benchmarking in Natural Language Understanding? arXiv preprint arXiv:2104.02145.

Chzhen, E.; Denis, C.; Hebiri, M.; and Lorieul, T. 2021. Set-valued classification–overview via a unified framework. arXiv preprint arXiv:2102.12318.

Denis, C.; and Hebiri, M. 2017. Confidence sets with expected sizes for multiclass classification. JMLR, 18(1): 3571–3598.

Devlin, J.; Chang, M.-W.; Lee, K.; and Toutanova, K. 2019. BERT: Pre-training of Deep Bidirectional Transformers for Language Understanding. In NAACL HLT, 4171–4186.

Ganea, O.-E.; and Hofmann, T. 2017. Deep Joint Entity Disambiguation with Local Neural Attention. In EMNLP, 2619–2629.

Gashteovski, K.; Gemulla, R.; Kotnis, B.; Hertling, S.; and Meilicke, C. 2020. On Aligning OpenIE Extractions with Knowledge Bases: A Case Study. In Proceedings of the First Workshop on Evaluation and Comparison of NLP Systems, 143–154.

Geva, M.; Goldberg, Y.; and Berant, J. 2019. Are We Modeling the Task or the Annotator? An Investigation of Annotator Bias in Natural Language Understanding Datasets. In EMNLP, 1161–1166.

Glockner, M.; Shwartz, V.; and Goldberg, Y. 2018. Breaking NLI Systems with Sentences that Require Simple Lexical Inferences. In ACL, 650–655.

Goel, K.; Rajani, N.; Vig, J.; Tan, S.; Wu, J.; Zheng, S.; Xiong, C.; Bansal, M.; and Ré, C. 2021. Robustness Gym: Unifying the NLP Evaluation Landscape. arXiv preprint arXiv:2101.04840.

Graham, Y.; Baldwin, T.; Moffat, A.; and Zobel, J. 2013. Continuous measurement scales in human evaluation of machine translation. In Proceedings of the 7th Linguistic Annotation Workshop and Interoperability with Discourse, 33–41.

Graham, Y.; Baldwin, T.; Moffat, A.; and Zobel, J. 2014. Is machine translation getting better over time? In EACL, 443–451.

Guo, Z.; and Barbosa, D. 2014. Robust entity linking via random walks. In CIKM, 499–508.

Gururangan, S.; Swayamdipta, S.; Levy, O.; Schwartz, R.; Bowman, S. R.; and Smith, N. A. 2018. Annotation artifacts in natural language inference data. arXiv preprint arXiv:1803.02324.

Hoffart, J.; Yosef, M. A.; Bordino, I.; Fürstenau, H.; Pinkal, M.; Spaniol, M.; Taneva, B.; Thater, S.; and Weikum, G. 2011. Robust disambiguation of named entities in text. In EMNLP, 782–792.

Hynes, N.; Sculley, D.; and Terry, M. 2017. The data linter: Lightweight, automated sanity checking for ml data sets. In NIPS MLSys Workshop.

Kiela, D.; Bartolo, M.; Nie, Y.; Kaushik, D.; Geiger, A.; Wu, Z.; Vidgen, B.; Prasad, G.; Singh, A.; Ringshia, P.; et al. 2021. Dynabench: Rethinking Benchmarking in NLP. arXiv preprint arXiv:2104.14337.

Kolitsas, N.; Ganea, O.-E.; and Hofmann, T. 2018. End-to-end neural entity linking. CoNLL.

Le, P.; and Titov, I. 2018. Improving Entity Linking by Modeling Latent Relations between Mentions. In ACL, 1595–1604.

Levy, O.; Goldberg, Y.; and Dagan, I. 2015. Improving distributional similarity with lessons learned from word embeddings. Transactions of the Association for Computational Linguistics, 3: 211–225.

Lin, C.-Y. 2004. Rouge: A package for automatic evaluation of summaries. In Text summarization branches out, 74–81.

Lundberg, S. M.; and Lee, S.-I. 2017. A Unified Approach to Interpreting Model Predictions. NeurIPS, 30: 4765–4774.

Maas, A.; Daly, R. E.; Pham, P. T.; Huang, D.; Ng, A. Y.; and Potts, C. 2011. Learning word vectors for sentiment analysis. In ACL, 142–150.

Northcutt, C. G.; Athalye, A.; and Mueller, J. 2021. Pervasive label errors in test sets destabilize machine learning benchmarks. arXiv preprint arXiv:2103.14749.

Oortwijn, Y.; Ossenkoppele, T.; and Betti, A. 2021. Interrater disagreement resolution: A systematic procedure to reach consensus in annotation tasks. In Proceedings of the Workshop on Human Evaluation of NLP Systems (HumEval), 131–141.

Papineni, K.; Roukos, S.; Ward, T.; and Zhu, W.-J. 2002. Bleu: a method for automatic evaluation of machine translation. In ACL, 311–318.

Park, C. W.; Jun, S. Y.; and MacInnis, D. J. 2000. Choosing what I want versus rejecting what I do not want: An application of decision framing to product option choice decisions. Journal of Marketing Research, 37(2): 187–202.

Paun, S.; Carpenter, B.; Chamberlain, J.; Hovy, D.; Kruschwitz, U.; and Poesio, M. 2018. Comparing bayesian models of annotation. TACL, 6: 571–585.

Poliak, A.; Naradowsky, J.; Haldar, A.; Rudinger, R.; and Van Durme, B. 2018. Hypothesis Only Baselines in Natural Language Inference. NAACL HLT, 180.

Prabhakaran, V.; Hutchinson, B.; and Mitchell, M. 2019. Perturbation Sensitivity Analysis to Detect Unintended Model Biases. In EMNLP, 5744–5749.

Radford, A.; Wu, J.; Child, R.; Luan, D.; Amodei, D.; and Sutskever, I. 2019. Language models are unsupervised multitask learners. OpenAI Blog, 1(8): 9.

Rajpurkar, P.; Jia, R.; and Liang, P. 2018. Know What You Don't Know: Unanswerable Questions for SQuAD. In ACL, 784–789.

Rajpurkar, P.; Zhang, J.; Lopyrev, K.; and Liang, P. 2016. Squad: 100,000+ questions for machine comprehension of text. arXiv preprint arXiv:1606.05250.

Resnick, P.; Kong, Y.; Schoenebeck, G.; and Weninger, T. 2021. Survey Equivalence: A Procedure for Measuring Classifier Accuracy Against Human Labels. arXiv preprint arXiv:2106.01254.

Ribeiro, M. T.; Guestrin, C.; and Singh, S. 2019. Are red roses red? evaluating consistency of question-answering models. In ACL, 6174–6184.

Ribeiro, M. T.; Wu, T.; Guestrin, C.; and Singh, S. 2020. Beyond Accuracy: Behavioral Testing of NLP Models with CheckList. In ACL, 4902–4912.

Richardson, M.; Burges, C. J.; and Renshaw, E. 2013. Mctest: A challenge dataset for the open-domain machine comprehension of text. In EMNLP, 193–203.

Rolnick, D.; Veit, A.; Belongie, S.; and Shavit, N. 2017. Deep learning is robust to massive label noise. arXiv preprint arXiv:1705.10694.

Rosales-Méndez, H.; Hogan, A.; and Poblete, B. 2019a. Fine-grained evaluation for entity linking. In EMNLP-IJCNLP, 718–727.

Rosales-Méndez, H.; Hogan, A.; and Poblete, B. 2019b. NIFify: Towards Better Quality Entity Linking Datasets. In WWW 2019, 815–818.

Sambasivan, N.; Kapania, S.; Highfill, H.; Akrong, D.; Paritosh, P.; and Aroyo, L. M. 2021. "Everyone wants to do the model work, not the data work": Data Cascades in High-Stakes AI. In CHI, 1–15.

Schwartz, R.; Sap, M.; Konstas, I.; Zilles, L.; Choi, Y.; and Smith, N. A. 2017. Story cloze task: Uw nlp system. In Proceedings of the 2nd Workshop on Linking Models of Lexical, Sentential and Discourse-level Semantics, 52–55.

Sevgili, O.; Shelmanov, A.; Arkhipov, M.; Panchenko, A.; and Biemann, C. 2020. Neural entity linking: A survey of models based on deep learning. arXiv preprint arXiv:2006.00575.

Sommerauer, P.; Fokkens, A.; and Vossen, P. 2020. Would you describe a leopard as yellow? Evaluating crowd-annotations with justified and informative disagreement. In COLING, 4798–4809.

Song, H.; Kim, M.; Park, D.; Shin, Y.; and Lee, J.-G. 2020. Learning from noisy labels with deep neural networks: A survey. arXiv preprint arXiv:2007.08199.

Trischler, A.; Wang, T.; Yuan, X.; Harris, J.; Sordoni, A.; Bachman, P.; and Suleman, K. 2016. Newsqa: A machine comprehension dataset. arXiv preprint arXiv:1611.09830.

Tsuchiya, M. 2018. Performance Impact Caused by Hidden Bias of Training Data for Recognizing Textual Entailment. In LREC.

van Hulst, J. M.; Hasibi, F.; Dercksen, K.; Balog, K.; and de Vries, A. P. 2020. REL: An entity linker standing on the shoulders of giants. In SIGIR, 2197–2200.

Zobel, J. 1998. How reliable are the results of large-scale information retrieval experiments? In SIGIR, 307–314.

Overconfidence in the Face of Ambiguity with Adversarial Data

Margaret Li* **and Julian Michael***

Paul G. Allen School of Computer Science & Engineering, University of Washington
{julianjm,margsli}@cs.washington.edu

Abstract

Adversarial data collection has shown promise as a method for building models which are more robust to the spurious correlations that generally appear in naturalistic data. However, adversarially-collected data may itself be subject to biases, particularly with regard to ambiguous or arguable labeling judgments. Searching for examples where an annotator disagrees with a model might over-sample ambiguous inputs, and filtering the results for high inter-annotator agreement may under-sample them. In either case, training a model on such data may produce predictable and unwanted biases. In this work, we investigate whether models trained on adversarially-collected data are miscalibrated with respect to the ambiguity of their inputs. Using Natural Language Inference models as a testbed, we find no clear difference in accuracy between naturalistically and adversarially trained models, but our model trained only on adversarially-sourced data is considerably more overconfident of its predictions and demonstrates worse calibration, especially on ambiguous inputs. This effect is mitigated, however, when naturalistic and adversarial training data are combined.

1 Introduction

End-to-end neural network models have had widespread success on standard benchmarks in NLP (Wang et al., 2018, 2019; Lee et al., 2017; Dozat and Manning, 2017). However, models trained with maximum-likelihood objectives under the standard Empirical Risk Minimization paradigm are liable to succeed in these settings by fitting to features or correlations in the data which are ultimately not representative of the underlying task and fail to generalize out of distribution, e.g., under domain shift or adversarial perturbation (Gururangan et al., 2018; Ilyas et al., 2019). One promising method to overcome this difficulty is to move past the ERM paradigm and learn or evaluate causal features which are invariant across domains or distributions of data. While methods to do this often require the use of explicitly specified domains of data (Peters et al., 2016; Arjovsky et al., 2020), a more lightweight approach is adversarial evaluation and training (Nie et al., 2020a; Kiela et al., 2021), in which annotators deliberately search for examples on which a model fails. Adversarial data annotation has been applied for a variety of tasks, including question answering (Bartolo et al., 2020), natural language inference (Nie et al., 2020a), hate speech detection (Vidgen et al., 2021), and sentiment analysis (Potts et al., 2021). Adversarial data can help reduce spurious correlations in existing data (Bartolo et al., 2020), expose a model's shortcomings in evaluation, and aid in training more robust models (Wallace et al., 2022).

However, the process of developing adversarial data is imperfect, and adversarial data may itself not resemble naturalistic distributions. For example, Phang et al. (2021) find that the AFLITE adversarial filtering algorithm (Sakaguchi et al., 2020; Bras et al., 2020), designed to find challenging examples in existing datasets, disproportionately favors contentious examples with annotator disagreement. This is suggestive that adversarially *collected* datasets, where humans actively try to fool a model, may be subject to these same biases Indeed, Phang et al. also show that adversarially-collected datasets may disproportionately penalize models that are similar to the one used during data collection. The qualitative properties of adversarially-collected data also vary depending on the adversary used during data collection, as shown by Williams et al. (2022) for the Adversarial NLI dataset (Nie et al., 2020a). For these reasons, it is not clear what a model's performance under adversarial evaluation implies about its performance characteristics on naturalistic distributions, nor is it clear how training on adversarial data aids a model's perfor-

*Equal contribution.

Proceedings of the First Workshop on Dynamic Adversarial Data Collection, pages 30 - 40
July 14, 2022 ©2022 Association for Computational Linguistics

mance in natural settings.

In this work, we focus on the interplay of adversarial learning and evaluation with *ambiguity*, or annotator disagreement. Just as adversarial filtering may over-sample ambiguous inputs (Phang et al., 2021), adversarial annotators may produce strange, ambiguous, or disputable inputs as they employ tricks to fool a model in the adversarial setting. To preempt this issue and ensure data quality, adversarial data collection methods filter out examples with low human agreement (Nie et al., 2020a), but it's possible that this approach could over-correct for the issue and *under*-sample such inputs in comparison to naturalistic data. For this reason, it is plausible that models trained on adversarially-collected data may be miscalibrated against the ambiguity of their inputs, forming a predictable blind spot.

We investigate this issue by training models on naturalistically and adversarially collected datasets, then comparing their performance with respect to gold annotator distributions. As a testbed, we use Natural Language Inference, an NLP benchmark task with already-available adversarial data (Nie et al., 2020a) and full annotator distributions (Nie et al., 2020b). We find no clear difference in accuracy between naturalistically and adversarially trained models, but our model trained only on adversarially-sourced data is considerably more overconfident of its predictions and demonstrates worse calibration, especially on ambiguous inputs. On the other hand, including both naturalistic data in training as well — as is standard practice (Nie et al., 2020a) — mitigates these issues. While our results do not raise alarms about standard practices with adversarial data, they suggest that we should keep in mind the importance of including naturalistic data in training regimes moving forward.[1]

2 Background: Robustness and Adversarial Data

Suppose we are interested in learning a conditional probability distribution $p(y \mid x)$. The classical machine learning approach of Empirical Risk Minimization does so with the use of input data drawn from a distribution D:

$$\operatorname*{argmin}_{\theta} \mathbb{E}_{x \sim D, y \sim p(\cdot|x)} - \log p(y|x, \theta), \quad (1)$$

where θ are the model parameters. However, this method can do a poor job of approximating $p(y \mid x)$ when x is drawn from very different distributions than D. One approach which has been used to address this is *robust optimization*, which minimizes the worst-case loss subject to some constraints (Madry et al., 2018; Ghaoui and Lebret, 1997; Wald, 1945). We can view robust optimization as solving a minimax problem:

$$\operatorname*{argmin}_{\theta} \max_{D \in \mathbb{D}} \mathbb{E}_{x \sim D, y \sim p(\cdot|x)} - \log p(y|x, \theta), \quad (2)$$

where $\mathbb{D}$ is a space of possible input distributions, and D is adversarially chosen among them. This formulation invites the question: what if $\mathbb{D}$ includes *all possible distributions?* Then we are free to find *any* x which the model gets wrong, and optimizing the loss effectively should produce a model which is robust to a wide range of distributions and hard to exploit.

This suggests a practical approach to improving robustness which involves actively searching for examples on which a model fails, and using those examples to train new, more robust models. This general approach has been applied in a variety of settings in NLP, such as the Build-It Break-It shared task (Ettinger et al., 2017), adversarial filtering of large datasets (Zellers et al., 2018; Sakaguchi et al., 2020), and adversarial benchmarking and leaderboards (Nie et al., 2020a; Kiela et al., 2021).

One complication that arises when sourcing adversarial data is with ambiguous or arguable examples. Suppose $\hat{\theta}$ perfectly models $p(y \mid x)$. Plugging this into Formula 2 yields $\max_{D \in \mathbb{D}} \mathrm{H}(Y \mid x)$, where D is concentrated on the inputs x which maximize the entropy of Y.

In this context, high entropy in the conditional distribution of Y corresponds to high *annotator disagreement*.[2] When a human searches for an adversarial example, they are looking for a *disagreement* between themselves and the model. In this setting, there may be competition for inclusion in these adversarial tasks between ambiguous examples on

[1]Code to reproduce our experiments is available at `https://github.com/julianmichael/aeae`.

[2]In this work, we assume all annotators implement the same probabilistic labeling function (which we are calling 'gold') and disagreement between annotators arises as an inherent feature of the task we are trying to model. We also assume that approximating annotator behavior on arguable or ambiguous examples is a desirable goal. These are simplifications: in some settings, e.g., the *prescriptive* paradigm of Röttger et al. (2022), we may wish to minimize annotator disagreement to learn a deterministic labeling function. In such settings, model behavior on arguable inputs may be uninteresting from the evaluation perspective, though searching for such examples could be useful for refining the task definition or annotation guidelines. We leave such issues out of scope for this work.

which the model is close to the gold (annotator) distribution and less ambiguous examples where the model is further from gold. Thus an adversarial data generation process may be biased towards input examples which are ambiguous but unhelpful for training.

Formally, a simple way to think about counteracting this may be to explicitly subtract the gold entropy from the loss being minimized:

$$\underset{\theta}{\operatorname{argmin}} \max_{D \in \mathbb{D}} \mathbb{E}_{x \sim D, y \sim p(\cdot|x)} - \log p(y \mid x, \theta) + \log p(y \mid x). \tag{3}$$

Here, the objective focuses the distribution D on examples which maximize the model's KL-Divergence from $p(y \mid x)$, no longer favoring ambiguous examples. Practical approaches to scaling adversarial data collection have applied a similar idea: in Adversarial NLI (Nie et al., 2020a) and Dynabench (Kiela et al., 2021), annotators are asked to find examples where they disagree with the model, and then these examples are only kept if multiple validators agree on the correct label. However, it is not clear how well-calibrated this process is: it might, for example, systematically omit genuinely ambiguous examples which the model gets wrong with high confidence. Whether training on data produced by this process results in pathological model behavior is what we test in this work.

3 Experimental Setup

Task Setting We use Natural Language Inference (Dagan et al., 2005; Bowman et al., 2015) as our underlying task, as there exist adversarial annotations for this task (Nie et al., 2020a; Kiela et al., 2021) and annotator disagreement has been well studied (Pavlick and Kwiatkowski, 2019; Nie et al., 2020b; Zhang and de Marneffe, 2021).

Model Variants We train models under three conditions:

- CLASSICAL: These models are trained on data elicited from annotators in a model-agnostic way, i.e., naturalistically.[3] For this we use the SNLI (Bowman et al., 2015) and MultiNLI (Williams et al., 2018) datasets.

[3]Unfortunately, since the NLI task is somewhat artificial, there is no "natural" distribution of input texts. This is one of the issues that leads to annotation artifacts in the first place (Gururangan et al., 2018) since some of the input text must be annotator-generated. Regardless, spurious correlations exist in any naturalistic distribution so we will use these training sets as proxies for something naturalistic.

Dataset	Train	Dev
SNLI	550,152	10,000
MultiNLI	392,702	10,000
ANLI (all rounds)	162,865	3,200
Chaos-SNLI		1,514
Chaos-MultiNLI		1,599

Table 1: Number of examples in training and development sets we use. For training data (top), development sets are used for model selection, while our evaluations (bottom) are on the ChaosNLI-annotated subsets of the SNLI and MultiNLI development sets.

- ADVERSARIAL: These models are trained on data elicited from annotators under the requirement that they must fool the model. For this we will use the adversarial annotations of Nie et al. (2020a).[4]
- ALL: These models are trained on the concatenation of all of the above data.

Evaluation Data We test the performance of our models in the setting where we have comprehensive distributions of annotator behavior. For this, we will use the ChaosNLI evaluation sets (Nie et al., 2020b) which have 100 independent annotations for each example (where the task is 3-way multi-class classification). ChaosNLI includes evaluation sets for SNLI (Bowman et al., 2015), MultiNLI (Williams et al., 2018), and αNLI (Bhagavatula et al., 2020, Abductive NLI). Of these, we use the SNLI and MultiNLI sets, since αNLI has a different task format than other NLI datasets. Dataset statistics are shown in Table 1.

Metrics Using densely-annotated evaluation data, we compute several evaluation metrics. Each metric is stratified across non-overlapping ranges of annotator agreement in order to analyze the dependence of model performance (or model differences) on the ambiguity of its input examples. Let $p(y)$ be the empirical distribution of annotator labels for an input example, and $\hat{y}$ be the model's prediction. Then, our metrics are:

- **Accuracy in Expectation:** The expectation of the accuracy of the model against a randomly sampled annotator in ChaosNLI (i.e.,

[4]In order for this to properly count as adversarial data for our model, we use the same model family as Nie et al. (2020a), which is BERT-large (Devlin et al., 2019) fine-tuned on SNLI and MultiNLI.

$p(\hat{y})$). We stratify this by the *human accuracy in expectation*, the accuracy of a randomly-sampled human against the plurality vote of all annotators ($\max_y(p(y))$). We use discrete bins to allow for precise comparison of model performance within and between different regimes of ambiguity.

- **Accuracy against Plurality:** The accuracy of the model against the plurality vote of the 100 annotators ($\hat{y} = \max(p(y))$). We also stratify this by human accuracy in expectation.
- **Model perplexity:** The exponentiated entropy of the model's predicted distribution; higher corresponds to more uncertainty. (This is independent of the gold labels.) We stratify this by the perplexity of the human annotator distribution.
- **KL-Divergence:** The KL-Divergence of the model's predicted label distribution against the empirical distribution of annotated labels. This gives a measure of how well-calibrated the model is with respect to the true annotator distribution. We stratify this measure by the entropy of the human annotator distribution.

Accuracy in expectation emulates the typical accuracy computation in an IID empirical risk minimization setting, while accuracy against plurality allows us to measure accuracy scores above human performance (assuming the plurality among 100 annotators can be treated as the ground truth).[5] We also include the annotator distribution as a human reference point (for KL-Divergence, this is 0 by construction).

Implementation Details

In all of our experiments, we begin with RoBERTa-Large (Liu et al., 2019), a masked language model pretrained on a large text corpus comprised of internet and book corpora. We then attach a classifier head and fine-tune each model according to the dataset combinations listed in Section 3. The model was implemented using the AllenNLP library and trained using the AdamW optimizer to maximize accuracy on the combined development sets of the model variant's respective corpora.

[5]The accuracy metrics provided for NLI datasets in practice are somewhere between the two, as the development and test sets of SNLI and MultiNLI were labeled by 5 annotators each and the majority label was chosen for the purposes of evaluation (Bowman et al., 2015; Williams et al., 2018).

4 Results

All results in this section are reported on the SNLI and MultiNLI development set portions of the ChaosNLI data. In all graphs, we provide smoothed kernel density estimates of the distributions over X and Y values in the margins where appropriate. Shaded areas around the lines represent 95% confidence intervals.

Accuracy in Expectation Model accuracy against randomly sampled annotators is shown in Figure 1. All models exhibit the same overall trend, approaching or reaching human performance on the most ambiguous and least ambiguous examples, with a dip in the middle of the range. Even if adversarial data collection does under-sample ambiguous inputs, we find no noticeable (or significant) effect on model performance in the low-agreement regime. A potential reason for this is that the baseline performance is already so low in these cases — very close to chance level — that there is little room for decreasing performance further.

Accuracy against Plurality Model accuracy against the plurality vote among annotators is shown in Figure 2. Once again, all models exhibit the same overall trend. While performance seems to level off or even increase for some models on extremely high-ambiguity examples (<50% human accuracy in expectation), there are too few such examples for us to draw any reliable conclusions in this regime.

Perplexity To understand the confidence levels of our models, we measure the perplexity of their output distributions and compare it to the perplexity of the human annotator distributions, shown in Figure 3. Here, there is a clear difference between ADVERSARIAL and the other models: it has extremely low perplexity on many more examples, and high perplexity on very few. Furthermore, while model perplexity is positively correlated with annotator perplexity for all models, the ADVERSARIAL model is less sensitive to it, with its perplexity growing less with respect to annotator perplexity. This suggests the adversarial data collection process may, on aggregate, favor examples with less ambiguity, skewing the behavior of the model. The ALL model, which was exposed to naturalistic data as well, does not display the same effect.

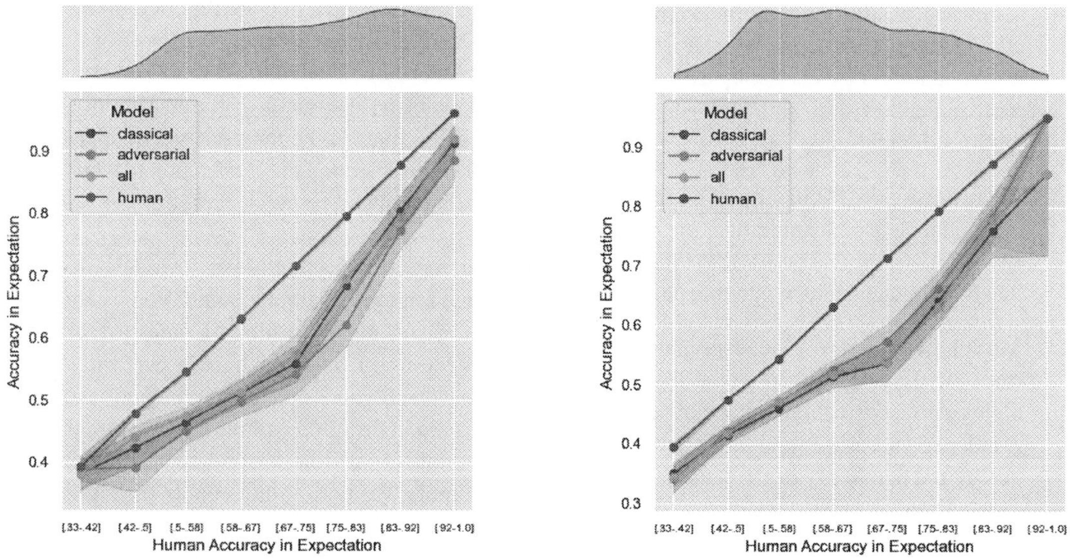

(a) Chaos-SNLI.

(b) Chaos-MultiNLI.

Figure 1: Model accuracy stratified by human accuracy, relative to a randomly sampled human judgment. Chance accuracy is approximately $\frac{1}{3}$, and the human baseline (which uses the plurality vote as the prediction) is an upper bound.

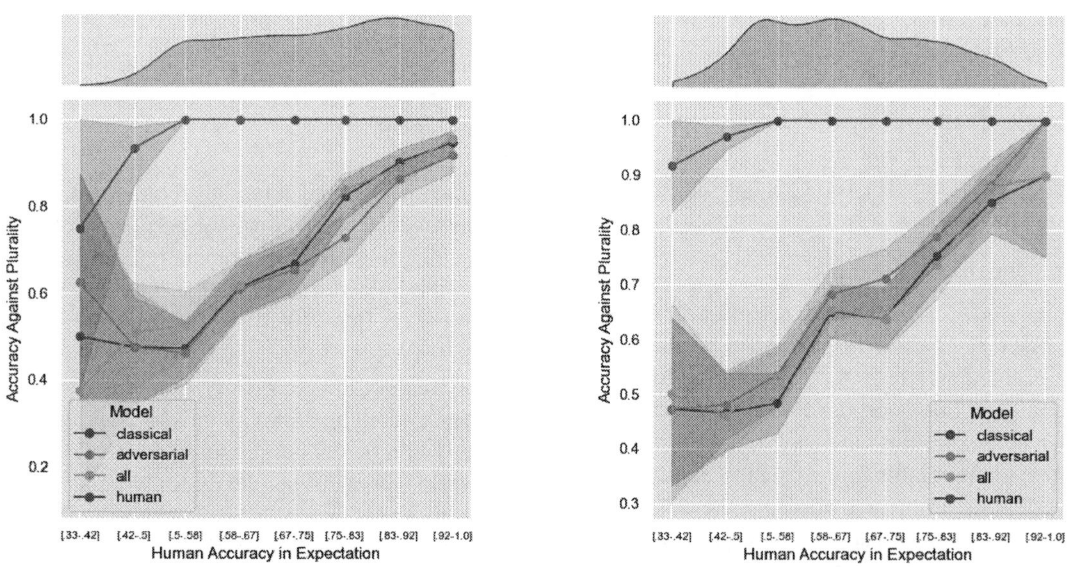

(a) Chaos-SNLI.

(b) Chaos-MultiNLI.

Figure 2: Model accuracy stratified by human accuracy, relative to the human plurality vote. The early dip in the human baseline below 50% is from a few cases with tied plurality votes (where we break ties randomly).

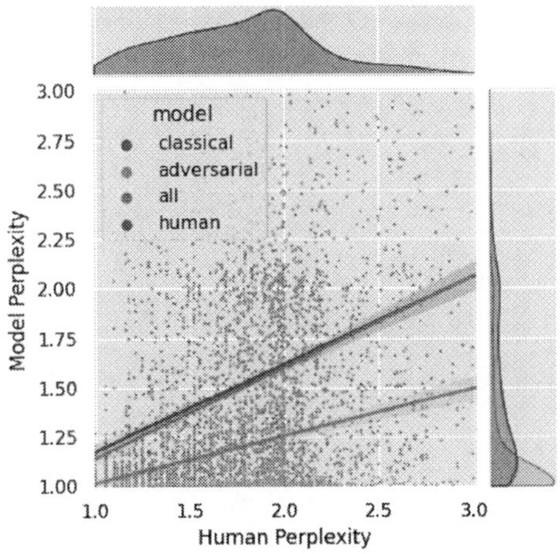

(a) Chaos-SNLI.

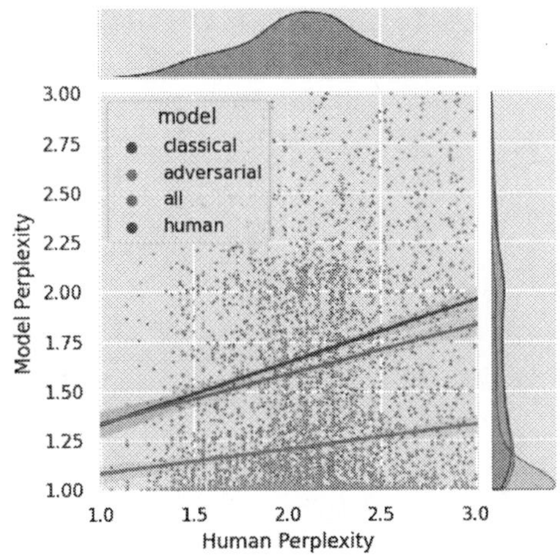

(b) Chaos-MultiNLI.

Figure 3: Model perplexity relative to annotator perplexity.

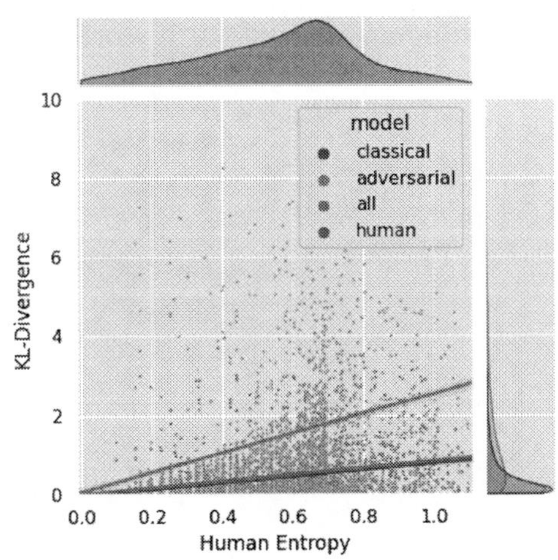

(a) Chaos-SNLI.

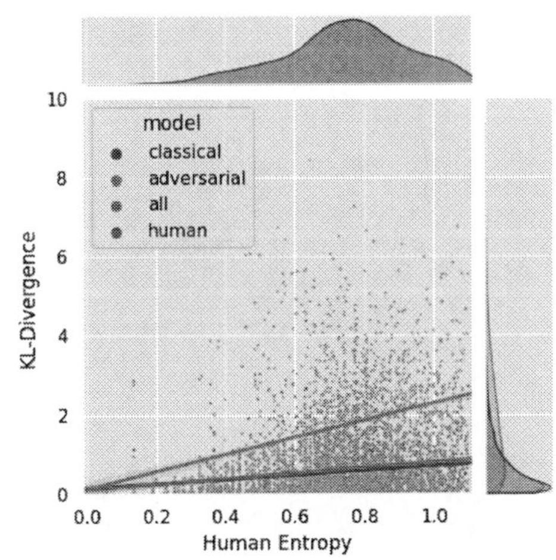

(b) Chaos-MultiNLI.

Figure 4: KL-Divergence of model outputs from the annotator distribution, graphed relative to annotator entropy. Both axes are measured in nats.

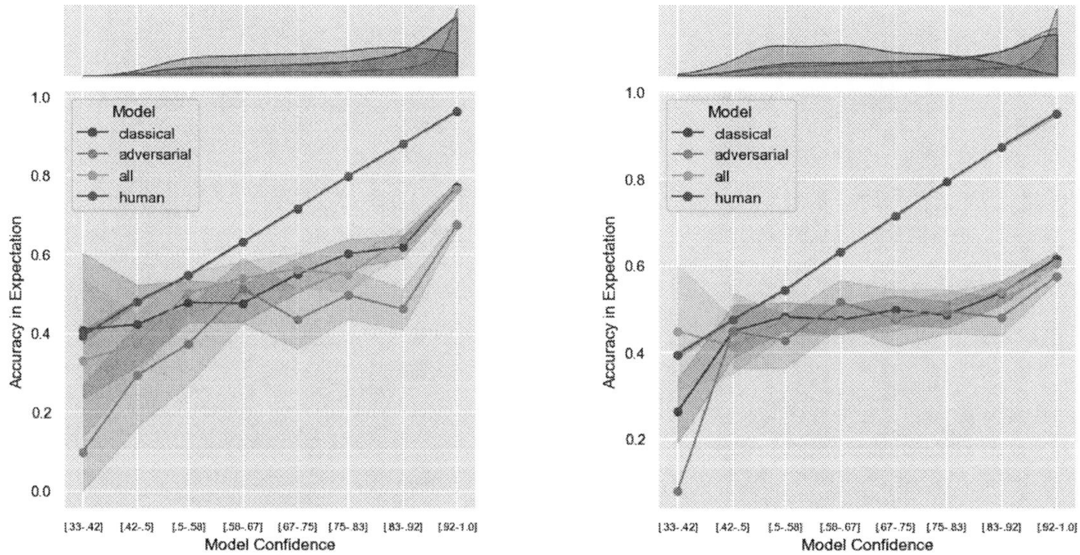

(a) Chaos-SNLI. (b) Chaos-MultiNLI.

Figure 5: Calibration curves for accuracy against a randomly sampled human. As the confidence score, we use the probability assigned by the model to its prediction.

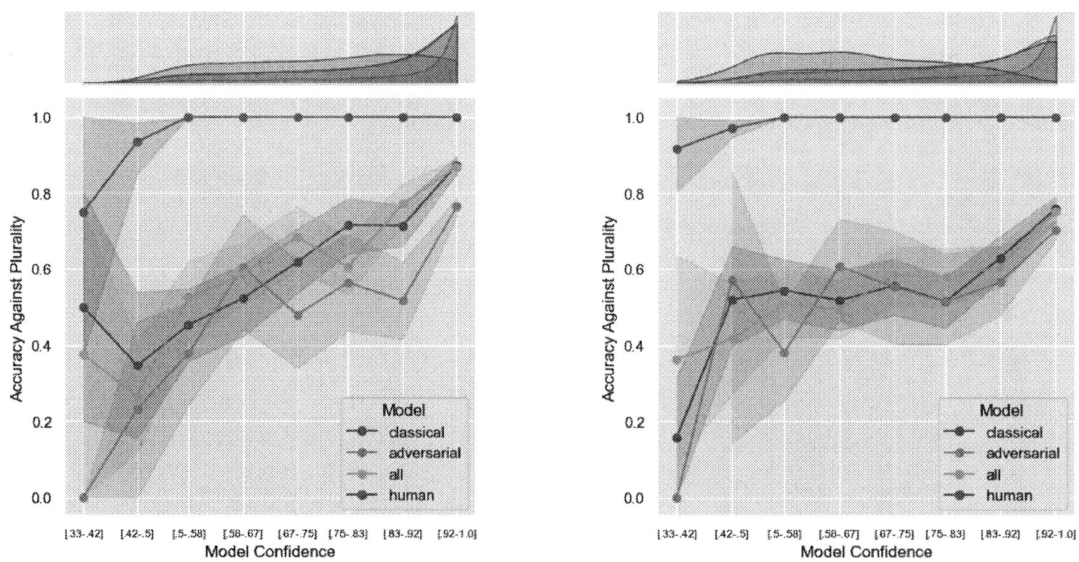

(a) Chaos-SNLI. (b) Chaos-MultiNLI.

Figure 6: Calibration curves for accuracy against the plurality vote among humans. As the confidence score, we use the probability assigned by the model to its prediction.

KL-Divergence To get a sense of how well the model fits the annotator distributions, we show the KL-Divergence of the models' predictions against the annotator distributions in Figure 4. What we find is that ADVERSARIAL diverges greatly from the gold distributions in comparison to CLASSICAL and ALL: it has much higher KL-Divergence in aggregate, many more examples with high KL-Divergence, and its KL-Divergence scores grow more quickly as the entropy of the annotator distribution increases. The biases in adversarial data collection, then, have led more to overconfidence on ambiguous examples than wrong predictions on unambiguous examples. These results provide supporting evidence for the hypothesis that training a model on adversarially-collected data may underexpose it to ambiguous examples and that this could have undesirable effects on its performance. However, these effects seem to be mitigated with the additional inclusion of naturalistic data (in the ALL model).

Calibration Calibration curves are shown in Figure 5. We find that the ADVERSARIAL model is highly confident more often than the other models, and at least in the very-high-confidence regime (>80% confidence), it has significantly worse calibration on SNLI (for MultiNLI, the results are borderline and only for the highest-confidence bin).

We also plot calibration curves relative to the plurality vote among annotators (Figure 6), which reflects the assumption that the model's maximum output probability reflects its epistemic uncertainty over the max-probability label. Here, the results are similar: the ADVERSARIAL model is worse calibrated in the very-high-confidence regime. Note, however, that when optimizing to maximize the likelihood of labels sampled from annotators, the output probabilities of a perfect model will not be well-calibrated against a plurality-based groundtruth. Optimizing for a model calibrated in this way is an alternative design choice which may require different training methods.

5 Discussion

In our experiments using SNLI, MultiNLI, and ANLI, we find that training only on adversarially-collected data produces similar accuracies across all regimes of ambiguity, but worse calibration at high confidence, and more overconfidence on ambiguous examples. This suggests that the adversarial data collection process may bias the model by favoring less ambiguous examples, but there are other potential interpretations of our results.

In particular, the observed miscalibration of ADVERSARIAL may the result of a more general domain shift between SNLI/MultiNLI and ANLI. This could explain why adding SNLI and MultiNLI to training, as in the ALL model, eliminates the effect. However, one might also expect to see a clear difference in accuracy as well if this were the issue. It's also worth noting that the SNLI and MNLI training sets are larger than ANLI's (see Table 1), which could explain why the ALL model behaves similarly to CLASSICAL. It remains an open question how little naturalistic (or, in-domain) data may be sufficient to mitigate the overconfidence issues we observe.

Some notable trends hold for all models we test. First, they all perform worse on ambiguous examples (Figure 1, Figure 2). This may be in part due to the relative scarcity of such examples in the training data or the relative difficulty of learning to model them. Second, they all demonstrate overconfidence, with model perplexity growing slower than human perplexity (Figure 3) and relatively poor calibration at high confidence levels (Figure 6). Even though augmenting training with adversarially-collected data has been shown to improve robustness in some settings (Bartolo et al., 2021a; Vidgen et al., 2021), our results do not yet show any benefits to calibration on ambiguous examples in existing data.

Finally, while we hypothesize that the overconfidence issue with training on adversarial data arises from filtering for annotator agreement, it is also possible that for ANLI, the adversarial annotators found examples that were less ambiguous in the first place (as annotators might, for example, want to focus on sure-fire model mistakes). Williams et al. (2022) found that about 5% of examples in ANLI "could reasonably be given multiple correct labels," suggesting a low level of ambiguity, but this was by the judgment of a single expert and may not correspond to the full variation in label assignment seen with crowdsourced annotators (which could potentially be investigated using the original unfiltered ANLI data). Measuring, controlling, managing, or representing ambiguity in adversarial annotation should be an interesting direction for future work, perhaps incorporating insights from recent work about construal (Trott et al., 2020; Pavlick and Kwiatkowski, 2019), explicit disambiguation (Min et al., 2020), model training dynamics (Swayamdipta et al., 2020; Liu et al., 2022), and other model-in-the-loop adversarial data collection efforts (Bartolo et al., 2020, 2021b; Vidgen et al., 2021; Potts et al., 2021).

6 Conclusion

We have shown that training only on adversarially-collected data, at least in the case of the Adversarial NLI (ANLI) dataset, can produce undesirable performance characteristics in the resulting models. In particular, when tested on SNLI and MultiNLI data, these models produce output distributions that are much further from annotator distributions and fail to accurately convey annotator uncertainty, with highly confident predictions even on highly ambiguous examples. It is also possible that adversarial training in this setting could produce lower prediction accuracy in regimes of low human agreement, but baseline accuracy is already so low for our models and data, and there are so few examples in the extremely-ambiguous regime, that such an effect is hard to find.

In our results, if a large amount of naturalistic data is also included in training (as in the ALL model) — as is standard practice — the overconfidence problem is mitigated. This is encouraging, as any adversarially-collected data must start with some naturalistic data to construct the initial adversary. However, it remains an open question how little naturalistic data is sufficient; a large enough seed corpus may be beneficial for avoiding such issues in a setting of dynamic adversarial data collection (Wallace et al., 2022). Future work can investigate this question, as well as how using full annotator distributions at training time (Zhang et al., 2021) or model calibration techniques may further help models deal with ambiguous inputs.

Acknowledgments

We would like to thank Ludwig Schmidt for early comments on this project, Ofir Press for providing entertainment, and the anonymous reviewers for their useful feedback.

References

Martin Arjovsky, Léon Bottou, Ishaan Gulrajani, and David Lopez-Paz. 2020. Invariant risk minimization.

Max Bartolo, Alastair Roberts, Johannes Welbl, Sebastian Riedel, and Pontus Stenetorp. 2020. Beat the AI: Investigating adversarial human annotation for reading comprehension. *Transactions of the Association for Computational Linguistics*, 8:662–678.

Max Bartolo, Tristan Thrush, Robin Jia, Sebastian Riedel, Pontus Stenetorp, and Douwe Kiela. 2021a. Improving question answering model robustness with synthetic adversarial data generation. In *Proceedings of the 2021 Conference on Empirical Methods in Natural Language Processing*, pages 8830–8848, Online and Punta Cana, Dominican Republic. Association for Computational Linguistics.

Max Bartolo, Tristan Thrush, Sebastian Riedel, Pontus Stenetorp, Robin Jia, and Douwe Kiela. 2021b. Models in the loop: Aiding crowdworkers with generative annotation assistants.

Chandra Bhagavatula, Ronan Le Bras, Chaitanya Malaviya, Keisuke Sakaguchi, Ari Holtzman, Hannah Rashkin, Doug Downey, Wen-tau Yih, and Yejin Choi. 2020. Abductive commonsense reasoning. In *8th International Conference on Learning Representations, ICLR 2020, Addis Ababa, Ethiopia, April 26-30, 2020*. OpenReview.net.

Samuel R. Bowman, Gabor Angeli, Christopher Potts, and Christopher D. Manning. 2015. A large annotated corpus for learning natural language inference. In *Proceedings of the 2015 Conference on Empirical Methods in Natural Language Processing (EMNLP)*. Association for Computational Linguistics.

Ronan Le Bras, Swabha Swayamdipta, Chandra Bhagavatula, Rowan Zellers, Matthew Peters, Ashish Sabharwal, and Yejin Choi. 2020. Adversarial filters of dataset biases. In *Proceedings of the 37th International Conference on Machine Learning*, volume 119 of *Proceedings of Machine Learning Research*, pages 1078–1088. PMLR.

Ido Dagan, Oren Glickman, and Bernardo Magnini. 2005. The pascal recognising textual entailment challenge. In *Proceedings of the First international conference on Machine Learning Challenges: evaluating Predictive Uncertainty Visual Object Classification, and Recognizing Textual Entailment*, pages 177–190.

Jacob Devlin, Ming-Wei Chang, Kenton Lee, and Kristina Toutanova. 2019. BERT: Pre-training of deep bidirectional transformers for language understanding. In *Proceedings of the 2019 Conference of the North American Chapter of the Association for Computational Linguistics: Human Language Technologies, Volume 1 (Long and Short Papers)*, pages 4171–4186, Minneapolis, Minnesota. Association for Computational Linguistics.

Timothy Dozat and Christopher D. Manning. 2017. Deep biaffine attention for neural dependency parsing. In *5th International Conference on Learning Representations, ICLR 2017, Toulon, France, April 24-26, 2017, Conference Track Proceedings*. OpenReview.net.

Allyson Ettinger, Sudha Rao, Hal Daumé III, and Emily M. Bender. 2017. Towards linguistically generalizable NLP systems: A workshop and shared task. In *Proceedings of the First Workshop on Building Linguistically Generalizable NLP Systems*, pages 1–10, Copenhagen, Denmark. Association for Computational Linguistics.

Laurent El Ghaoui and Hervé Lebret. 1997. Robust solutions to least-squares problems with uncertain data. *SIAM J. Matrix Anal. Appl.*, 18(4):1035–1064.

Suchin Gururangan, Swabha Swayamdipta, Omer Levy, Roy Schwartz, Samuel Bowman, and Noah A. Smith. 2018. Annotation artifacts in natural language inference data. In *Proceedings of the 2018 Conference of the North American Chapter of the Association for Computational Linguistics: Human Language Technologies, Volume 2 (Short Papers)*, volume 2, pages 107–112. Association for Computational Linguistics.

Andrew Ilyas, Shibani Santurkar, Dimitris Tsipras, Logan Engstrom, Brandon Tran, and Aleksander Madry. 2019. Adversarial examples are not bugs, they are features. In *Advances in Neural Information Processing Systems*, volume 32. Curran Associates, Inc.

Douwe Kiela, Max Bartolo, Yixin Nie, Divyansh Kaushik, Atticus Geiger, Zhengxuan Wu, Bertie Vidgen, Grusha Prasad, Amanpreet Singh, Pratik Ringshia, Zhiyi Ma, Tristan Thrush, Sebastian Riedel, Zeerak Waseem, Pontus Stenetorp, Robin Jia, Mohit Bansal, Christopher Potts, and Adina Williams. 2021. Dynabench: Rethinking benchmarking in NLP. *CoRR*, abs/2104.14337.

Kenton Lee, Luheng He, Mike Lewis, and Luke Zettlemoyer. 2017. End-to-end neural coreference resolution. In *Proceedings of the 2017 Conference on Empirical Methods in Natural Language Processing*, pages 188–197. Association for Computational Linguistics.

Alisa Liu, Swabha Swayamdipta, Noah A. Smith, and Yejin Choi. 2022. Wanli: Worker and ai collaboration for natural language inference dataset creation.

Yinhan Liu, Myle Ott, Naman Goyal, Jingfei Du, Mandar Joshi, Danqi Chen, Omer Levy, Mike Lewis, Luke Zettlemoyer, and Veselin Stoyanov. 2019. Roberta: A robustly optimized bert pretraining approach. *arXiv preprint arXiv:1907.11692*.

Aleksander Madry, Aleksandar Makelov, Ludwig Schmidt, Dimitris Tsipras, and Adrian Vladu. 2018. Towards deep learning models resistant to adversarial attacks. In *6th International Conference on Learning Representations, ICLR 2018, Vancouver, BC, Canada, April 30 - May 3, 2018, Conference Track Proceedings*. OpenReview.net.

Sewon Min, Julian Michael, Hannaneh Hajishirzi, and Luke Zettlemoyer. 2020. AmbigQA: Answering ambiguous open-domain questions. In *Proceedings of the 2020 Conference on Empirical Methods in Natural Language Processing (EMNLP)*, pages 5783–5797, Online. Association for Computational Linguistics.

Yixin Nie, Adina Williams, Emily Dinan, Mohit Bansal, Jason Weston, and Douwe Kiela. 2020a. Adversarial NLI: A new benchmark for natural language understanding. In *Proceedings of the 58th Annual Meeting of the Association for Computational Linguistics*, pages 4885–4901, Online. Association for Computational Linguistics.

Yixin Nie, Xiang Zhou, and Mohit Bansal. 2020b. What can we learn from collective human opinions on natural language inference data? In *Proceedings of the 2020 Conference on Empirical Methods in Natural Language Processing (EMNLP)*, pages 9131–9143, Online. Association for Computational Linguistics.

Ellie Pavlick and Tom Kwiatkowski. 2019. Inherent disagreements in human textual inferences. *Transactions of the Association for Computational Linguistics*, 7:677–694.

Jonas Peters, Peter Bühlmann, and Nicolai Meinshausen. 2016. Causal inference by using invariant prediction: identification and confidence intervals. *Journal of the Royal Statistical Society: Series B (Statistical Methodology)*, 78(5):947–1012.

Jason Phang, Angelica Chen, William Huang, and Samuel R. Bowman. 2021. Adversarially constructed evaluation sets are more challenging, but may not be fair. *CoRR*, abs/2111.08181.

Christopher Potts, Zhengxuan Wu, Atticus Geiger, and Douwe Kiela. 2021. DynaSent: A dynamic benchmark for sentiment analysis. In *Proceedings of the 59th Annual Meeting of the Association for Computational Linguistics and the 11th International Joint Conference on Natural Language Processing (Volume 1: Long Papers)*, pages 2388–2404, Online. Association for Computational Linguistics.

Paul Röttger, Bertie Vidgen, Dirk Hovy, and Janet B. Pierrehumbert. 2022. Two contrasting data annotation paradigms for subjective nlp tasks. In *Proceedings of the 2022 Conference of the North American Chapter of the Association for Computational Linguistics: Human Language Technologies*, Seattle, WA. Association for Computational Linguistics.

Keisuke Sakaguchi, Ronan Le Bras, Chandra Bhagavatula, and Yejin Choi. 2020. Winogrande: An adversarial winograd schema challenge at scale. *Proceedings of the AAAI Conference on Artificial Intelligence*, 34(05):8732–8740.

Swabha Swayamdipta, Roy Schwartz, Nicholas Lourie, Yizhong Wang, Hannaneh Hajishirzi, Noah A. Smith, and Yejin Choi. 2020. Dataset cartography: Mapping and diagnosing datasets with training dynamics. In *Proceedings of the 2020 Conference on Empirical Methods in Natural Language Processing (EMNLP)*, pages 9275–9293, Online. Association for Computational Linguistics.

Sean Trott, Tiago Timponi Torrent, Nancy Chang, and Nathan Schneider. 2020. (Re)construing meaning in NLP. In *Proceedings of the 58th Annual Meeting of the Association for Computational Linguistics*, pages 5170–5184, Online. Association for Computational Linguistics.

Bertie Vidgen, Tristan Thrush, Zeerak Waseem, and Douwe Kiela. 2021. Learning from the worst: Dynamically generated datasets to improve online hate detection. In *Proceedings of the 59th Annual Meeting of the Association for Computational Linguistics and the 11th International Joint Conference on Natural Language Processing (Volume 1: Long Papers)*, pages 1667–1682, Online. Association for Computational Linguistics.

Abraham Wald. 1945. Statistical decision functions which minimize the maximum risk. *Annals of Mathematics*, 45(2):265–280.

Eric Wallace, Adina Williams, Robin Jia, and Douwe Kiela. 2022. Analyzing dynamic adversarial training data in the limit. In *Findings of the Association for Computational Linguistics*.

Alex Wang, Yada Pruksachatkun, Nikita Nangia, Amanpreet Singh, Julian Michael, Felix Hill, Omer Levy, and Samuel R. Bowman. 2019. SuperGLUE: A multi-task benchmark and analysis platform for natural language understanding. In H. Wallach, H. Larochelle, A. Beygelzimer, F. d'Alché-Buc, E. Fox, and R. Garnett, editors, *Advances in Neural Information Processing Systems 32*, pages 3261–3275. Curran Associates, Inc.

Alex Wang, Amanpreet Singh, Julian Michael, Felix Hill, Omer Levy, and Samuel Bowman. 2018. GLUE: A multi-task benchmark and analysis platform for natural language understanding. In *Proceedings of the 2018 EMNLP Workshop BlackboxNLP: Analyzing and Interpreting Neural Networks for NLP*, pages 353–355. Association for Computational Linguistics.

Adina Williams, Nikita Nangia, and Samuel Bowman. 2018. A broad-coverage challenge corpus for sentence understanding through inference. In *Proceedings of the 2018 Conference of the North American Chapter of the Association for Computational Linguistics: Human Language Technologies, Volume 1 (Long Papers)*, pages 1112–1122, New Orleans, Louisiana. Association for Computational Linguistics.

Adina Williams, Tristan Thrush, and Douwe Kiela. 2022. ANLIzing the adversarial natural language inference dataset. In *Proceedings of the Society for Computation in Linguistics 2022*, pages 23–54, online. Association for Computational Linguistics.

Rowan Zellers, Yonatan Bisk, Roy Schwartz, and Yejin Choi. 2018. SWAG: A large-scale adversarial dataset for grounded commonsense inference. In *Proceedings of the 2018 Conference on Empirical Methods in Natural Language Processing*, pages 93–104, Brussels, Belgium. Association for Computational Linguistics.

Shujian Zhang, Chengyue Gong, and Eunsol Choi. 2021. Learning with different amounts of annotation: From zero to many labels. In *Proceedings of the 2021 Conference on Empirical Methods in Natural Language Processing*, pages 7620–7632, Online and Punta Cana, Dominican Republic. Association for Computational Linguistics.

Xinliang Frederick Zhang and Marie-Catherine de Marneffe. 2021. Identifying inherent disagreement in natural language inference. In *Proceedings of the 2021 Conference of the North American Chapter of the Association for Computational Linguistics: Human Language Technologies*, pages 4908–4915, Online. Association for Computational Linguistics.

longhorns at DADC 2022: How many linguists does it take to fool a Question Answering model? A systematic approach to adversarial attacks.

Venelin Kovatchev[1]†* Trina Chatterjee[3]* Venkata S Govindarajan[2]*
Jifan Chen[3] Eunsol Choi[3] Gabriella Chronis[2] Anubrata Das[1] Katrin Erk[2]
Matthew Lease[1] Junyi Jessy Li[2] Yating Wu[4] Kyle Mahowald[2]*
* Contributors towards the official submission
[1] School of Information, The University of Texas at Austin
[2] Department of Linguistics, The University of Texas at Austin
[3] Department of Computer Science, The University of Texas at Austin
[4] Department of Electrical and Computer Engineering, The University of Texas at Austin

Abstract

Developing methods to adversarially challenge NLP systems is a promising avenue for improving both model performance and interpretability. Here, we describe the approach of the team "longhorns" on Task 1 of the The First Workshop on Dynamic Adversarial Data Collection (DADC), which asked teams to manually fool a model on an Extractive Question Answering task. Our team finished first, with a model error rate of 62%.[1] We advocate for a systematic, linguistically informed approach to formulating adversarial questions, and we describe the results of our pilot experiments, as well as our official submission.

1 Introduction

Rapid progress in NLP has resulted in systems obtaining apparently super-human performance on popular benchmarks such as GLUE (Wang et al., 2018), SQUaD (Rajpurkar et al., 2016), and SNLI (Bowman et al., 2015). Dynabench (Kiela et al., 2021) proposes an alternative approach to benchmarking: a dynamic benchmark wherein a human adversary creates examples that can "fool" a state-of-the-art model but not a human language user. The idea is that, by generating and compiling examples that fool a particular system, the community can gain a better idea of that system's actual strengths and weaknesses, as well as ideas and data for iteratively improving it.

There is no straightforward recipe, however, for generating successful adversarial examples. To contribute to that knowledge base, this paper describes the strategy used by team "longhorns" in Task 1 of The First Workshop on Dynamic Adversarial Data Collection (DADC), which was on Extractive Question Answering (answering a question about a passage by pointing to a particular span of text within that passage).[2] We focus not only on describing the details of our strategy, but also on our process for approaching the task. At the time of this paper submission, pending expert validation of the results, our team ranked first in the competition, obtaining 62% Model Error Rate (MER).

Our approach towards creating adversarial examples was designed to be systematic, analytical, and draw on linguistically informed ideas. We first compiled a list of linguistically inspired "attack strategies" and used it to create adversarial examples in a systematic manner. We then analyzed some existing biases of the model-in-the-loop and its performance on a variety of different attacks. We used this piloting phase to select the best performing attacks for the official submission.

Based on the approaches that were most successful both in our pilot studies and in our official submission, we posit that the following broad areas should be of particular interest for theoretically motivated adversarial attacks on contemporary NLP systems, as evidenced by their strong performance on our target task:

- **Taking advantage of models' strong priors.** The model was proficient at identifying the correct kind of named entity being asked for (e.g., a person for a "who" question, a place for a "where" question), but was biased to give answers which were salient (either topically or because they appeared first (Ko et al., 2020)) or which had high lexical overlap with the question. Thus, picking a distractor with the same entity type as the target answer (e.g., another person mentioned in the text when

†Primary author and coordinator (venelin@utexas.edu)

[1]The results and the team ranking are pending validation from the organizers of the task at the time of the submission.

[2]https://dadcworkshop.github.io/shared-task/

Proceedings of the First Workshop on Dynamic Adversarial Data Collection, pages 41 - 52
July 14, 2022 ©2022 Association for Computational Linguistics

the question was a "who" question) was often effective. This result is broadly consistent with observations that modern NLP systems can perform well in the general case but can be biased towards frequency-based priors (e.g., Wei et al., 2021) that mean they are sometimes "right for the wrong reasons" (McCoy et al., 2019).

- **Using language that is linguistically taxing for humans (and machines) to process.** Psycholinguists who study human language processing often study constructions that are grammatical but difficult for humans to process in real time, such as garden path sentences (Frazier and Rayner, 1982; Ferreira and Henderson, 1991) and complex coreference resolution (Kaiser and Fedele, 2019; Durrett and Klein, 2013). We found that the model was indeed often fooled by questions that included these types of constructions. While we did not collect any human data, the sentences that fooled the model are likely to be hard for humans as measured by tests of real-time processing difficulty (e.g., eye tracking, self-paced reading), even though humans would be able to successfully process these sentences given enough time.

- **Tapping into domain-general, non-linguistic reasoning.** We found that asking questions which do not require mere linguistic processing but require other kinds of reasoning (e.g., numerical reasoning, temporal reasoning, common-sense reasoning, list manipulation) were hard for the model. This result is consistent with prior work showing that language models struggle with these kinds of reasoning tasks (Marcus, 2020; Elazar et al., 2021; Talmor et al., 2020) and may be more generally explained by evidence from cognitive science that these kinds of reasoning tap into cognitive processes that are distinct from linguistic processing (Diachek et al., 2020; Blank et al., 2014).

Because these strategies and this general approach are broad and theoretically motivated, we believe that our methods could be used to generate adversarial examples on other Natural Language Understanding tasks besides Question Answering. In what follows, we characterize our approach in both the pilot phase and official submission, provide our list of attack strategies, and discuss the limitations of the task and model.

2 Task Definition

In Task 1 of DADC, titled "Better Annotators", each participating team submits 100 "official" extractive question answering (QA) examples through the Dynabench platform. The organizers of the shared task provide short passages as context and the participants have to create questions that "can be correctly answered from a span in the passage and DO NOT require a Yes or No answer". The objective is to find as many model-fooling examples as possible – the winning team is the one with the highest validated model error rate (vMER)[3].

The competition uses Dynabench (Kiela et al., 2021): "an open-source platform for dynamic dataset creation and model benchmarking". Dynabench aims to facilitate human-and-model-in-the-loop dataset creation. The annotators' aim to generate examples that will be misclassified by an automated model, but can be answered correctly by competent human speakers. Dynabench has been used to create data for Question Answering (Kaushik et al., 2021), Natural Language Inference (Williams et al., 2022), Online Hate Detection (Vidgen et al., 2021), and Sentiment Analysis (Potts et al., 2021), among others.

3 Approach

Our team consisted of faculty, postdocs, and students from the UT Austin linguistics department, computer science department, information school, and electrical and computer engineering department.

We approached the problem of creating adversarial attacks in a systematic manner, informed by ideas from computational linguistics, psycholinguistics, and theoretical linguistics. We composed a list of linguistic phenomena and reasoning capabilities that we hypothesized would be difficult for a state-of-the-art QA model. We then used some of those phenomena to create our official submission of adversarial examples. While the list is not exhaustive, it covers a wide range of potential attack strategies and can be used to guide the creation of adversarial examples for other tasks and systems.

[3]For full instructions, see `https://dadcworkshop.github.io/shared-task/`

Separate from our official submission for the competition, we ran a series of pilot experiments in which we used the list as a guide for experimenting with a variety of strategies for creating adversarial example. To ensure a fair and competitive official submission, all pilot experiments were carried out either before the official start of the shared task or after the official submission was made.

Our objective when evaluating the different adversarial strategies was to explore the space of potential attack strategies to determine the most successful ones for fooling the model. For each strategy, we measured the Model Error Rate (MER) on a small sample of example, and we also analyzed how frequently the attack can be used.

Based on the results of these pilot experiments, we targeted the best strategies for our official submission. Official question submissions were made by subsets of the team, in group sizes ranging from 1 to around 10. Since not all strategies can be used for all example passages, we used the specific passages we were presented with in order to guide our decision about what strategy to focus on for a particular question. When more than one participant was present, question submissions were made by consensus agreement among those present.

Anecdotally, we found that the attacks were often more successful when multiple team members are present, with each member hypothesizing model behaviors from diverse angles. Overall, we found the adversarial question generation process nontrivial, taking 5-10 minutes per passage, although we became faster over time. We also chose to skip passages occasionally when the passage covers very well-known entity, is too simple, or is not promising to most of our strategies (not having distractor entities, etc). We generated multiple questions for promising passages.

4 Pilot Experiments: Evaluating Adversarial Strategies

Many of our "adversarial strategies" are inspired by prior work in adversarial data generation and unit testing for Question Answering, Natural Language Inference, and Paraphrase Identification (Glockner et al., 2018; Kovatchev et al., 2018; Naik et al., 2018; Dua et al., 2019; Kovatchev et al., 2019; Nie et al., 2019; Wallace et al., 2019; Bartolo et al., 2020; Gardner et al., 2020; Hossain et al., 2020; Jeretic et al., 2020; Kaushik et al., 2020; Ribeiro et al., 2020; Saha et al., 2020). We propose the following linguistic and reasoning phenomena as a source for potential adversarial attacks:

Lexical knowledge Examples that require understanding lexical properties and in partiuclar lexical entailments that require knowledge of **hypernyms/hyponyms** (e.g., knowing that dog => animal, but animal =/> dog), **named entities** and their properties (e.g., knowing Shakira is a singer), **nominalization** (e.g., knowing "a submission" implies something has been submitted), **(a)symmetrical relations** (e.g., knowing that John marrying Mary implies Mary marrying John, but John loving Mary does not imply Mary loving John), **polarity substitutions** (e.g., knowing that a movie is good implies that it is not bad), **converse substitution** (e.g., knowing that if something has been provided, it has been received), **comparisons with antonyms** (e.g., knowing that if Clara is the tallest, she is not shorter than Mary), **reasoning about modal verbs** (e.g., understanding that if something *could* happen, that does not mean it *did* happen), and **reasoning about quantifers** (e.g., knowing that if some swans are white, that does not imply all swans are white).

Syntax and Discourse knowledge Examples that require syntactic or discourse-level understanding such as **Genitives** (e.g., knowing that elephant's foot = the foot of the elephant) and **Datives** (e.g., knowing that give her a cake = give a cake to her). **Relative Clauses** can be used in attacks to either include distracting information (e.g., "Maria, who is the president of the company" when the correct answer has nothing to do with Maria's role in the company) or to specify the correct referent (e.g., "the actor who bought the house" when that actor must be distinguished from a set of other actors).

When **Conjunction** or **Disjunction** appear in the passage (e.g., John and Mary love strawberries and cake, but John doesn't like chocolate), an adversarial question targets the ability of the model to correctly identify the syntactic scope (e.g.: Who loves cake and chocolate?). Closely related are the phenomena of **Intersectivity** (e.g., knowing that "a singer and a good man" =/> a good singer) and **Restrictivity** (e.g., understanding that "all my work due today" =/> all my work).

When a complex **prepositional phrase attachment** appears in the passage (e.g., "I saw two men with a telescope in the park"), an adversarial question requires disambiguation (e.g., "Who has the telescope"). Questions based on the **Argumenta-**

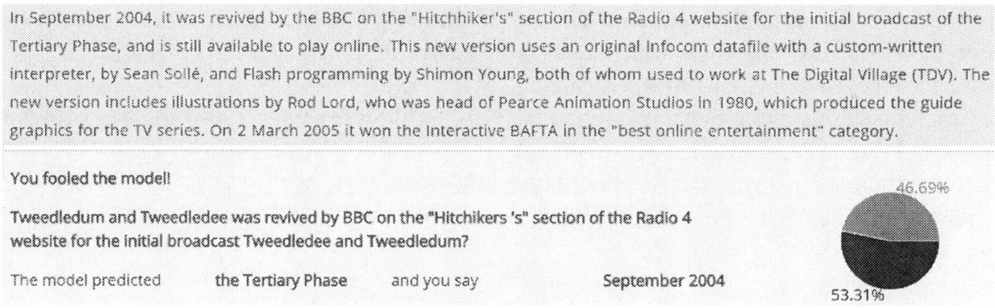

Figure 1: An example of "semantic similarity" model bias: when the model is fed a nonsensical question, it responds with an answer with high semantic overlap with the question.

tive structure require the model to correctly identify the core arguments (e.g., knowing that "John broke the vase" implies that "The vase broke"; but does not imply "John broke."). This attack can be further complicated when the same verb appears multiple times in the passage. Adversarial attacks based on **Ellipsis**, **Anaphora**, and **Coreference** test the ability of the model to process long distance syntactic dependencies.

Negation can appear both in the passage and in the question. It can be expressed in a variety of ways: simple (e.g., no, not), adverbial (e.g., never), pronoun (e.g., nobody), morphological (e.g., unfinished), lexical (e.g., refuse to), implicit (e.g., I wish I had a boat), double negation. Adversarial questions can also target the ability of the model to identify the scope of negation either in the question or in the passage.

Garden Path questions (e.g., Who is the director of the movie directing?) are syntactically confusing and much-studied in psycholinguistics for causing processing difficulty in humans (Frazier, 1979).

Questions about **Mental States** of individuals are inspired by work in psychology showing that it can be challenging to reason about the mental states of others (e.g., knowing that "Why does Maria think that Sandra is leaving?" could require a different answer than "Why is Sandra leaving?"(Wellman, 1992; Kovatchev et al., 2020).

Reasoning Questions that require various kinds of non-linguistic reasoning such as **Conditionals and hypothetical situations** (e.g., Who would be the champion if Mary didn't lose the final?), **Numerical Reasoning** (e.g., Who is the second richest person?), **Temporal Reasoning** (e.g., What happened in a specific timeframe?), **Commonsense reasoning** (including logical implications, contradiction, etc.), and **List manipulations** (e.g., Which two of the actors in the list are male?).

Finally, **distractor**-based attacks make use of model priors by expanding the question with additional information. **Meaningful distractors** directs the model towards a wrong answer, while the strategy of **adding noise** relies on increasing the complexity of the question.

The different phenomena can appear in the passages, in the question, or in both. Not every phenomena can be used to generate attacks for every passage, and the phenomena also appear with different frequency in the data. In our pilot experiments we distributed the different phenomena across the members of the team. We measured the Model Error Rate (MER) for the different strategies and determined how frequently each attack could be used.

5 Exploring Model Biases

During the pilot experiment step in Section 4 we found that the model-in-the-loop performs surprisingly well on a variety of different attacks. We hypothesized that at least in some situations, the strong performance is due to spurious correlations, the nature of the underlying language model, and the nature of the task. We further carried out a set of experiments to determine some specifics of the model behavior. We briefly discuss two "shortcuts" used by the model.

Semantic similarity Figure 1 illustrates the model bias towards "semantic similarity" on a nonsensical question. When the model is unsure what to do, or like in this example, when the question is not a valid English sentence, it identifies parts of the context that are similar to the question and predicts neighboring words. Due to the relatively short length of most of the passages, this strategy

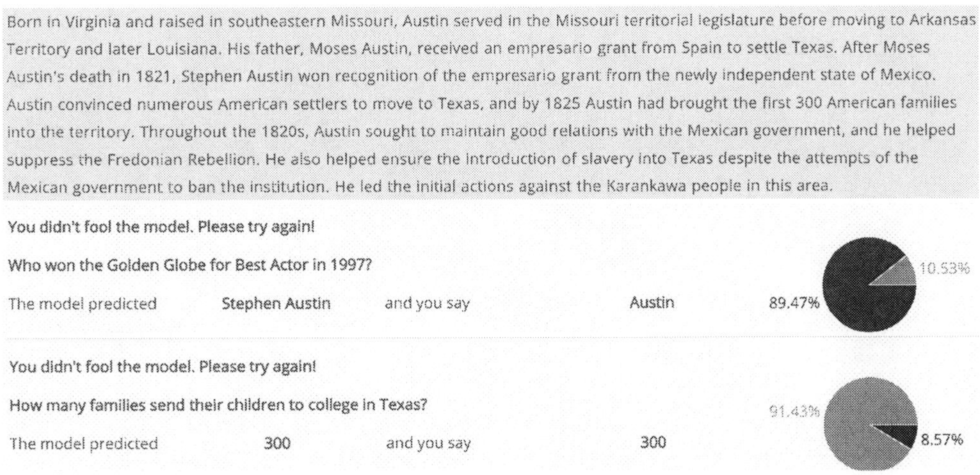

Figure 2: When faced with nonsensical questions, the model will give salient answers of the correct entity type. The passage is about Texas colonizer Stephen Austin, but the first question is about the Golden Globe awards. The model confidently answers that Stephen Austin won the Gold Globe for Best Actor in 1997.

often gets the correct answer without actually understanding the question.

Type of question We noticed that the model is very good at identifying some properties of the answer based on the type of question. For example, a "who" question typically asks for a named entity, while a "how many" question asks for a quantity. A strong heuristic adopted by the model is to return an answer of the correct "type" regardless of the actual question. Figure 2 illustrates that: neither question can be answered from the passage, but the model makes a guess based on the type of question. Once again, the short length of the passages and the fact that they typically contain just a few tokens of the correct "type" artificially boosts the performance of the model.

In our official submission, we used those model biases to increase the difficulty of the adversarial examples. When possible, we used those biases to guide the model towards a wrong answer. In passages where we could not confuse the model (e.g.: only one named entity in a "who" question), we rephrased the questions in such a way that makes it harder for the model to use heuristics.

6 Official Submission

After analyzing and discussing the results of our preliminary experiments, for our official submission we focus on the following strategies: using distractors, numerical reasoning, temporal reasoning, garden path questions, complex coreference, list manipulations, and common-sense reasoning. We also used the model biases to either confuse the model or reduce it's ability to rely on heuristics. In the rest of this section we briefly describe each of our strategies and provide examples.

6.1 Taking advantage of model priors

Distractors One of the most successful and easy-to-use adversarial strategies was using distractors. An example of that strategy can be seen in Figure 3: the phrasing of question has a high degree of similarity with the portion of the text talking about narrow belts ("between X and Y AU and relatively sharp boundaries"), however the correct response is "wide belts" due to the specified sizes. The "distractor" strategy can be used frequently. We often combined the distractor strategies with other strategies. For example, in Figure 3, we combine it with "numerical reasoning". Anecdotally, we found the distractors to be most successful when the correct answer was not the most salient entity of its type in the passage (e.g., targeting a briefly mentioned director in a passage mostly about one particular actor) and when there were many other entities of the desired type available, as opposed to just 1 or 2 (e.g., a "who" question for a passage that mentions 10 people is more challenging than a "who" question for a passage that mentions only 1 person).

6.2 Linguistically difficult utterances

Garden Path Questions Figure 4 shows an example of a garden path question (Frazier, 1979).

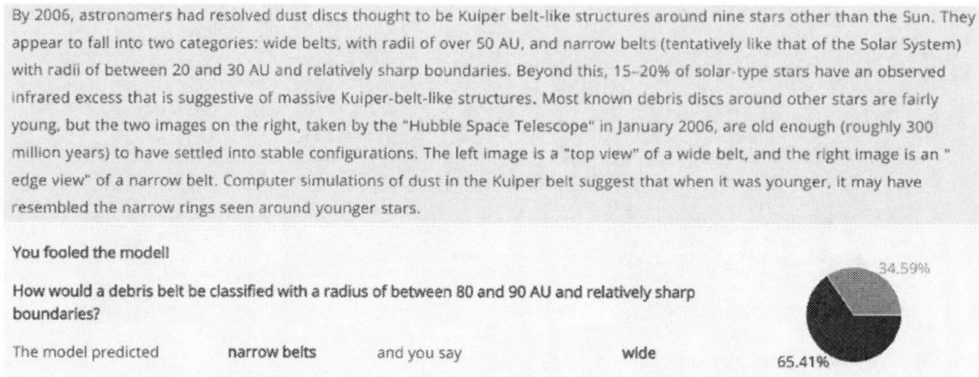

Figure 3: An example of "distractor" and "numerical reasoning" strategies. The model has to reason that "between 80 and 90" is greater than the 50 AU boundary identified in the passage.

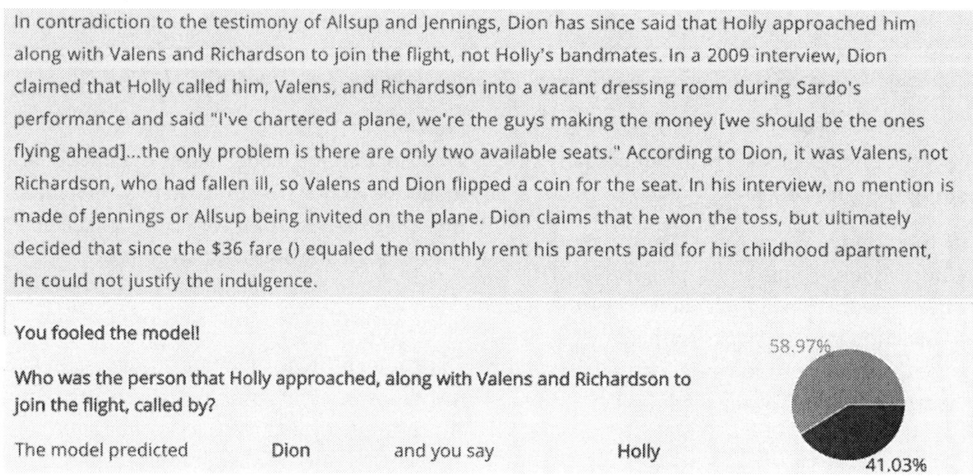

Figure 4: "Garden path" strategy. Until the very end of the sentence, the question seems to be about something else.

Until the last word, the question appears to be asking about "the person that Holly approached, along with Valens and Richardson" (to which the answer would be Dion). But the last word makes it clear that the reader needs to reparse the question, to see that it is actually asking who *calls* that person (i.e., Dion) – which makes the answer Holly. The model is unable to correctly process the complex syntactic structure of the sentence and responds "Dion". Garden path questions are easier to generate than temporal and numerical reasoning questions since they can be generated for a wider variety of texts, but we found that the model can often handle even quite complex syntactic constructions. We hypothesize that this is mainly due to the length of the passages and the "type of question" model bias.

Anaphora and Coreference Adversarial examples based on anaphora and coreference can require the model to demonstrate the ability to resolve long distance syntactic dependencies and often require making common-sense inferences as well. In Figure 5, the founders were worried about their own death. To correctly respond to the question, the model first has to identify "their" as the answer and then resolve the coreference between "their" and "the founders". Instead, the model just returns a salient named entity. Examples based on anaphora and coreference are relatively infrequent, as they require multiple entities and potentially ambiguous coreference in the passage.

6.3 Non-linguistic reasoning

Numerical Reasoning Adversarial examples based on numerical reasoning require the model to carry out simple mathematical calculations or comparisons to identify the correct answer. For example, in Figure 3, the model had to calculate

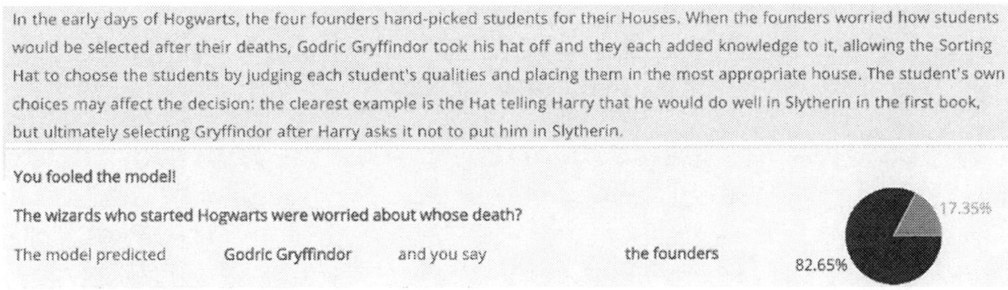

Figure 5: "Coreference" strategy. The model has to figure out that the word "their" refers to the founders.

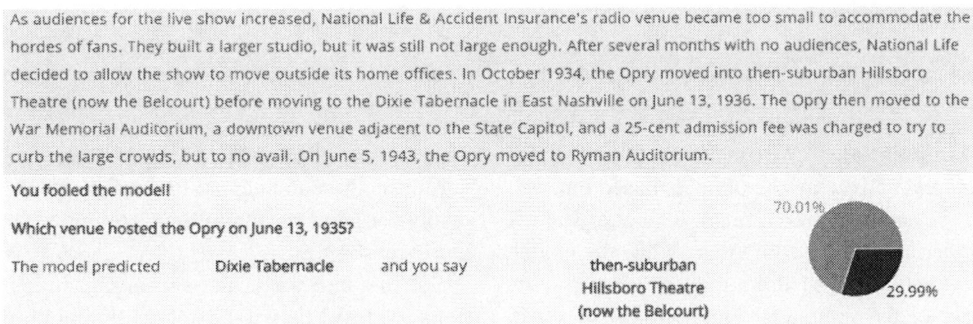

Figure 6: "Temporal reasoning" strategy. The model has to understand that June 13, 1935 is during the period when the Opry moved to the Hillsboro Theatre.

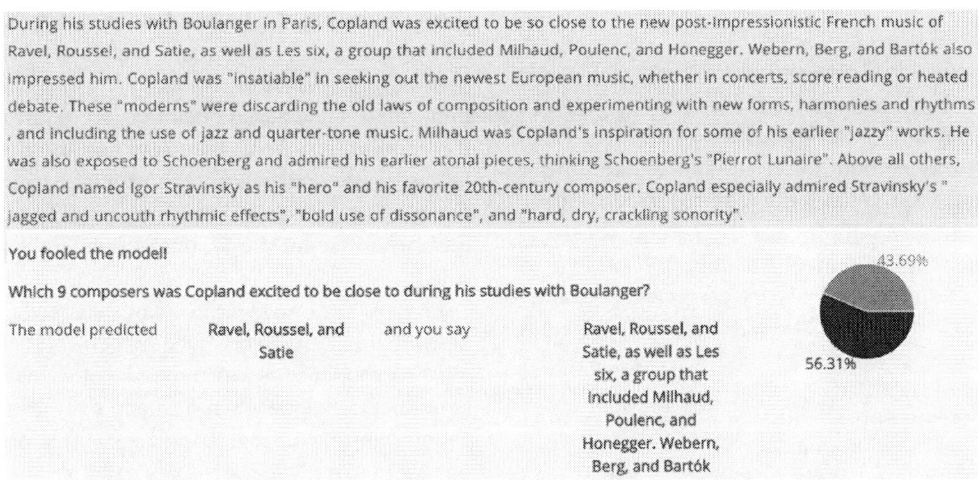

Figure 7: "List manipulations" strategy. The model has to identify the 9 composers asked for, but only gives 3.

that "between 80 and 90" is "over 50" in order to answer correctly.

Temporal Reasoning Adversarial examples based on temporal reasoning require the model to reason about the chronological order of events and the different states of the world at different points in time. In the example shown in Figure 6 the Opry moves to Hillsboro in 1934 and then to Dixie Tabernacle in 1936. We ask the model for a date that is not mentioned explicitly (June 13, 1935). The correct answer is "Hillsboro", but the model is fooled by recognizing a portion of the date (June 13) and predicts Dixie Tabernacle.

Temporal-based examples are relatively rare, as they require the passage to have multiple dates as well as multiple different events and world states associated with the dates. However, when available, temporal-based attacks were very successful.

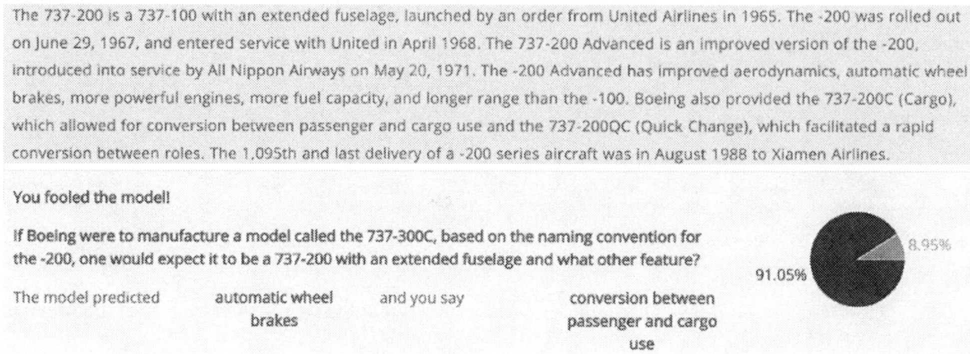

Figure 8: "Common-sense reasoning" strategy. The model has to flexibly adapt the naming convention to a hypothetical example: a kind of creative reasoning task that humans do easily but that models often struggle with.

List Manipulations We used two different strategies to create adversarial examples based on lists. Figure 7 illustrates one of them. The question requests the full list of 9 composers, while the model only extracts the first three due to the syntactic structure of the sentence. The second list-based strategy asks for a subset of a list that fulfills certain criteria. List-based adversarial attacks are relatively infrequent in a single passage setting we study.

Common-sense Reasoning Adversarial attacks based on common sense reasoning test the basic understanding of the world of the model or its ability to reason about different entities and objects. In Figure 8, the model is required to break apart the string "737-300C" and "737-200C" correctly and then reason about the naming convention: we are told that "C" stands for cargo and so the hypothetical "737-300C" should also have the cargo feature.

7 Discussion

A fundamental feature of language is that it is a cooperative enterprise (Clark, 1996) that enables efficient communication between parties (Gibson et al., 2019). Therefore, in ordinary language, people typically talk about discourse-relevant entities (Sperber and Wilson, 1986), avoid difficult syntactic constructions (Futrell et al., 2015; Gibson, 1998), and structure information in a way that is easy to produce and understand (MacDonald, 2013; Levy, 2008). If anything unifies all of our most successful attack strategies, it is that they eschew these principles in the context of the given task and passages. Instead, successful attacks ask about surprising aspects of the text (e.g., by including distractors), often using complex language (e.g., garden path sentences and complex coreference resolution) and reasoning (e.g., temporal and numeric reasoning).

So, in some ways, the successful attack questions are less likely to be encountered in ordinary language use (leading to claims that adversarial examples are brittle, e.g., Phang et al., 2021; Bowman, 2022). But another key property of human language is that it is flexible and generative, such that people can produce and understand surprising and unexpected utterances. To that end, we think these adversarial questions are a fair target for improving systems precisely because they are linguistically unusual: human language is not just for the "average case" but can be used to express meanings that are subtle, interesting, and complicated.

Perhaps because these questions also require humans to think creatively outside their ordinary linguistic experience, we also found that we achieved better performance when we had larger groups of people working on generating questions at once, so that there was a wider diversity of ideas.

Indeed, while some questions may be less likely to appear in a "extractive question answering" dataset, they are understandable by humans and are likely to be useful for efficient communication in real-world settings. The objective behind "extractive QA" is that a machine should answer any question that a human would, given the passage. A variety of real-world tasks can be reduced to extractive QA and in many cases the "correct" passage corresponding to the question is not known a priori. Asking questions such as "Where was X at a time Y" and "What is the difference between 737-200 and 737-200C" may be less natural for a human that has access to the passage, but are questions that someone would, for example, ask their automated assistant. Therefore, a well functioning model needs to embrace the creativity and be able to correctly answer adversarial questions.

Finally, the adversarial attacks that we present are not just interesting from the scientific point of view, but also have clear practical implications. Most of the attacks correspond to specific capacities of the model-in-the-loop such as coreference resolution, numerical and temporal reasoning. The consistently high MER indicates that the model underperforms in tasks that require those capacities.

Our approach towards creating adversarial examples allows us to implicitly evaluate the performance of the model and the quality of the data with respect to a wide variety of linguistic and reasoning categories. Overall, we found that the model-in-the-loop performs impressively good on the majority of question types. Only a small subset of the strategies could consistently obtain above 50% MER and these strategies did not necessarily work for all questions. For instance, questions with relatively few possible entities matching the question type meant fewer possibilities for distractors.

The performance of the model is also a function of the varying difficulty of the passages. We found the majority of the passages to be short declarative texts with a simple syntactic structure, few named entities, and low amount of information. Generating and answering questions from those passages is a rather trivial task. The selection of more complex paragraphs will likely result in a lower performance of the model and a lot more possibilities for creative and successful adversarial attacks.

8 Conclusions

In this paper we presented the strategies used by team "longhorns" for Task 1 of DADC: generating high-quality adversarial examples. We obtain the best results in the competition by taking a systematic approach, using linguistic knowledge, and working in a collaborative environment.

Our approach outperforms prior work in terms of model error rate and also provides a variety of insights. For instance, our pilot analysis covers a large number of linguistic and reasoning phenomena and explores different model biases. This facilitates a more in-depth analysis of the performance of the model. The systematic approach also gives us insight into the quality and difficulty of the data.

Our strategies for generating adversarial examples are not limited to extractive question answering. They can be adopted at larger scale to improve the quality of models and data on a variety of different tasks. We believe that our work opens new research directions with both scientific and practical implications.

Acknowledgements

This research was supported in part by NSF grants IIS-1850153 and IIS-2107524, as well as by Wipro, the Knight Foundation, the Micron Foundation, and by Good Systems,[4] a UT Austin Grand Challenge to develop responsible AI technologies. The statements made herein are solely the opinions of the authors and do not reflect the views of the sponsoring agencies.

References

Max Bartolo, Alastair Roberts, Johannes Welbl, Sebastian Riedel, and Pontus Stenetorp. 2020. Beat the AI: investigating adversarial human annotations for reading comprehension. *CoRR*, abs/2002.00293.

Idan Blank, Nancy Kanwisher, and Evelina Fedorenko. 2014. A functional dissociation between language and multiple-demand systems revealed in patterns of bold signal fluctuations. *Journal of neurophysiology*, 112(5):1105–1118.

Samuel Bowman. 2022. The dangers of underclaiming: Reasons for caution when reporting how NLP systems fail. In *Proceedings of the 60th Annual Meeting of the Association for Computational Linguistics (Volume 1: Long Papers)*, pages 7484–7499, Dublin, Ireland. Association for Computational Linguistics.

Samuel R. Bowman, Gabor Angeli, Christopher Potts, and Christopher D. Manning. 2015. A large annotated corpus for learning natural language inference. In *Proceedings of the 2015 Conference on Empirical Methods in Natural Language Processing (EMNLP)*. Association for Computational Linguistics.

Herbert H Clark. 1996. *Using language*. Cambridge university press.

Evgeniia Diachek, Idan Blank, Matthew Siegelman, Josef Affourtit, and Evelina Fedorenko. 2020. The domain-general multiple demand (md) network does not support core aspects of language comprehension: a large-scale fmri investigation. *Journal of Neuroscience*, 40(23):4536–4550.

Dheeru Dua, Yizhong Wang, Pradeep Dasigi, Gabriel Stanovsky, Sameer Singh, and Matt Gardner. 2019.

[4] `http://goodsystems.utexas.edu/`

DROP: A reading comprehension benchmark requiring discrete reasoning over paragraphs. In *Proceedings of the 2019 Conference of the North American Chapter of the Association for Computational Linguistics: Human Language Technologies, Volume 1 (Long and Short Papers)*, pages 2368–2378, Minneapolis, Minnesota. Association for Computational Linguistics.

Greg Durrett and Dan Klein. 2013. Easy victories and uphill battles in coreference resolution. In *Proceedings of the 2013 Conference on Empirical Methods in Natural Language Processing*, pages 1971–1982, Seattle, Washington, USA. Association for Computational Linguistics.

Yanai Elazar, Hongming Zhang, Yoav Goldberg, and Dan Roth. 2021. Back to square one: Artifact detection, training and commonsense disentanglement in the Winograd schema. In *Proceedings of the 2021 Conference on Empirical Methods in Natural Language Processing*, pages 10486–10500, Online and Punta Cana, Dominican Republic. Association for Computational Linguistics.

Fernanda Ferreira and John M Henderson. 1991. Recovery from misanalyses of garden-path sentences. *Journal of Memory and Language*, 30(6):725–745.

Lyn Frazier and Keith Rayner. 1982. Making and correcting errors during sentence comprehension: Eye movements in the analysis of structurally ambiguous sentences. *Cognitive psychology*, 14(2):178–210.

Lynn Frazier. 1979. On comprehending sentences: Syntactic parsing strategies. *ETD Collection for University of Connecticut*.

Richard Futrell, Kyle Mahowald, and Edward Gibson. 2015. Large-scale evidence of dependency length minimization in 37 languages. *Proceedings of the National Academy of Sciences*, 112(33):10336–10341.

Matt Gardner, Yoav Artzi, Victoria Basmov, Jonathan Berant, Ben Bogin, Sihao Chen, Pradeep Dasigi, Dheeru Dua, Yanai Elazar, Ananth Gottumukkala, Nitish Gupta, Hannaneh Hajishirzi, Gabriel Ilharco, Daniel Khashabi, Kevin Lin, Jiangming Liu, Nelson F Liu, Phoebe Mulcaire, Qiang Ning, Sameer Singh, Noah A Smith, Sanjay Subramanian, Reut Tsarfaty, Eric Wallace, Ally Zhang, and Ben Zhou. 2020. Evaluating models' local decision boundaries via contrast sets.

Edward Gibson. 1998. Linguistic complexity: Locality of syntactic dependencies. *Cognition*, 68(1):1–76.

Edward Gibson, Richard Futrell, Steven P Piantadosi, Isabelle Dautriche, Kyle Mahowald, Leon Bergen, and Roger Levy. 2019. How efficiency shapes human language. *Trends in cognitive sciences*, 23(5):389–407.

Max Glockner, Vered Shwartz, and Yoav Goldberg. 2018. Breaking NLI systems with sentences that require simple lexical inferences. In *Proceedings of the 56th Annual Meeting of the Association for Computational Linguistics (Volume 2: Short Papers)*, pages 650–655, Melbourne, Australia. Association for Computational Linguistics.

Md Mosharaf Hossain, Venelin Kovatchev, Pranoy Dutta, Tiffany Kao, Elizabeth Wei, and Eduardo Blanco. 2020. An analysis of natural language inference benchmarks through the lens of negation. In *Proceedings of the 2020 Conference on Empirical Methods in Natural Language Processing (EMNLP)*, pages 9106–9118, Online. Association for Computational Linguistics.

Paloma Jeretic, Alex Warstadt, Suvrat Bhooshan, and Adina Williams. 2020. Are natural language inference models IMPPRESsive? Learning IMPlicature and PRESupposition. In *Proceedings of the 58th Annual Meeting of the Association for Computational Linguistics*, pages 8690–8705, Online. Association for Computational Linguistics.

Elsi Kaiser and Emily Fedele. 2019. Reference resolution. *The Oxford Handbook of Reference*.

Divyansh Kaushik, Eduard Hovy, and Zachary Lipton. 2020. Learning the difference that makes a difference with counterfactually-augmented data. In *International Conference on Learning Representations*.

Divyansh Kaushik, Douwe Kiela, Zachary C. Lipton, and Wen-tau Yih. 2021. On the efficacy of adversarial data collection for question answering: Results from a large-scale randomized study. In *Proceedings of the 59th Annual Meeting of the Association for Computational Linguistics and the 11th International Joint Conference on Natural Language Processing (Volume 1: Long Papers)*, pages 6618–6633, Online. Association for Computational Linguistics.

Douwe Kiela, Max Bartolo, Yixin Nie, Divyansh Kaushik, Atticus Geiger, Zhengxuan Wu, Bertie Vidgen, Grusha Prasad, Amanpreet Singh, Pratik Ringshia, Zhiyi Ma, Tristan Thrush, Sebastian Riedel, Zeerak Waseem, Pontus Stenetorp, Robin Jia, Mohit Bansal, Christopher Potts, and Adina Williams. 2021. Dynabench: Rethinking benchmarking in NLP. In *Proceedings of the 2021 Conference of the North American Chapter of the Association for Computational Linguistics: Human Language Technologies*, pages 4110–4124, Online. Association for Computational Linguistics.

Miyoung Ko, Jinhyuk Lee, Hyunjae Kim, Gangwoo Kim, and Jaewoo Kang. 2020. Look at the first sentence: Position bias in question answering. *ArXiv*, abs/2004.14602.

Venelin Kovatchev, M. Antònia Martí, and Maria Salamó. 2018. ETPC - a paraphrase identification corpus annotated with extended paraphrase typology and negation. In *Proceedings of the Eleventh International Conference on Language Resources and Evaluation (LREC 2018)*, Miyazaki, Japan. European Language Resources Association (ELRA).

Venelin Kovatchev, M. Antonia Marti, Maria Salamo, and Javier Beltran. 2019. A qualitative evaluation framework for paraphrase identification. In *Proceedings of the International Conference on Recent Advances in Natural Language Processing (RANLP 2019)*, pages 568–577, Varna, Bulgaria. INCOMA Ltd.

Venelin Kovatchev, Phillip Smith, Mark Lee, Imogen Grumley Traynor, Irene Luque Aguilera, and Rory Devine. 2020. "what is on your mind?" automated scoring of mindreading in childhood and early adolescence. In *Proceedings of the 28th International Conference on Computational Linguistics*, pages 6217–6228, Barcelona, Spain (Online). International Committee on Computational Linguistics.

Roger Levy. 2008. Expectation-based syntactic comprehension. *Cognition*, 106(3):1126–1177.

Maryellen C. MacDonald. 2013. How language production shapes language form and comprehension. *Frontiers in Psychology*, 4:226.

Gary Marcus. 2020. The next decade in ai: four steps towards robust artificial intelligence. *arXiv preprint arXiv:2002.06177*.

Tom McCoy, Ellie Pavlick, and Tal Linzen. 2019. Right for the wrong reasons: Diagnosing syntactic heuristics in natural language inference. In *Proceedings of the 57th Annual Meeting of the Association for Computational Linguistics*, pages 3428–3448, Florence, Italy. Association for Computational Linguistics.

Aakanksha Naik, Abhilasha Ravichander, Norman Sadeh, Carolyn Rose, and Graham Neubig. 2018. Stress test evaluation for natural language inference. In *Proceedings of the 27th International Conference on Computational Linguistics*, pages 2340–2353, Santa Fe, New Mexico, USA. Association for Computational Linguistics.

Yixin Nie, Yicheng Wang, and Mohit Bansal. 2019. Analyzing compositionality-sensitivity of nli models. *Proceedings of the AAAI Conference on Artificial Intelligence*, 33(01):6867–6874.

Jason Phang, Angelica Chen, William Huang, and Samuel R Bowman. 2021. Adversarially constructed evaluation sets are more challenging, but may not be fair. *arXiv preprint arXiv:2111.08181*.

Christopher Potts, Zhengxuan Wu, Atticus Geiger, and Douwe Kiela. 2021. DynaSent: A dynamic benchmark for sentiment analysis. In *Proceedings of the 59th Annual Meeting of the Association for Computational Linguistics and the 11th International Joint Conference on Natural Language Processing (Volume 1: Long Papers)*, pages 2388–2404, Online. Association for Computational Linguistics.

Pranav Rajpurkar, Jian Zhang, Konstantin Lopyrev, and Percy Liang. 2016. SQuAD: 100,000+ questions for machine comprehension of text. In *Proceedings of the 2016 Conference on Empirical Methods in Natural Language Processing*, pages 2383–2392, Austin, Texas. Association for Computational Linguistics.

Marco Tulio Ribeiro, Tongshuang Wu, Carlos Guestrin, and Sameer Singh. 2020. Beyond accuracy: Behavioral testing of NLP models with CheckList. In *Proceedings of the 58th Annual Meeting of the Association for Computational Linguistics*, pages 4902–4912, Online. Association for Computational Linguistics.

Swarnadeep Saha, Yixin Nie, and Mohit Bansal. 2020. ConjNLI: Natural language inference over conjunctive sentences. In *Proceedings of the 2020 Conference on Empirical Methods in Natural Language Processing (EMNLP)*, pages 8240–8252, Online. Association for Computational Linguistics.

Dan Sperber and Deirdre Wilson. 1986. *Relevance: Communication and cognition*, volume 142. Citeseer.

Alon Talmor, Yanai Elazar, Yoav Goldberg, and Jonathan Berant. 2020. olmpics-on what language model pre-training captures. *Transactions of the Association for Computational Linguistics*, 8:743–758.

Bertie Vidgen, Tristan Thrush, Zeerak Waseem, and Douwe Kiela. 2021. Learning from the worst: Dynamically generated datasets to improve online hate detection. In *Proceedings of the 59th Annual Meeting of the Association for Computational Linguistics and the 11th International Joint Conference on Natural Language Processing (Volume 1: Long Papers)*, pages 1667–1682, Online. Association for Computational Linguistics.

Eric Wallace, Pedro Rodriguez, Shi Feng, Ikuya Yamada, and Jordan Boyd-Graber. 2019. Trick me if you can: Human-in-the-loop generation of adversarial examples for question answering. *Transactions of the Association for Computational Linguistics*, 7:387–401.

Alex Wang, Amanpreet Singh, Julian Michael, Felix Hill, Omer Levy, and Samuel Bowman. 2018. GLUE: A multi-task benchmark and analysis platform for natural language understanding. In *Proceedings of the 2018 EMNLP Workshop BlackboxNLP: Analyzing and Interpreting Neural Networks for NLP*, pages 353–355, Brussels, Belgium. Association for Computational Linguistics.

Jason Wei, Dan Garrette, Tal Linzen, and Ellie Pavlick. 2021. Frequency effects on syntactic rule learning in transformers. In *Proceedings of the 2021 Conference on Empirical Methods in Natural Language Processing*, pages 932–948, Online and Punta Cana, Dominican Republic. Association for Computational Linguistics.

Henry M Wellman. 1992. *The child's theory of mind.* The MIT Press.

Adina Williams, Tristan Thrush, and Douwe Kiela. 2022. ANLIzing the adversarial natural language inference dataset. In *Proceedings of the Society for Computation in Linguistics 2022*, pages 23–54, online. Association for Computational Linguistics.

Collecting high-quality adversarial data for machine reading comprehension tasks with humans and models in the loop

Damian Y. Romero Diaz,[♣,✳] **Magdalena Anioł,**[✳] **John Culnan**[♣]
damian@explosion.ai*, magda@explosion.ai, jmculnan@arizona.edu
[✳] Explosion, [♣] University of Arizona

Abstract

We present our experience as annotators in the creation of high-quality, adversarial machine-reading-comprehension data for extractive QA for Task 1 of the First Workshop on Dynamic Adversarial Data Collection (DADC). DADC is an emergent data collection paradigm with both models and humans in the loop. We set up a quasi-experimental annotation design and perform quantitative analyses across groups with different numbers of annotators focusing on successful adversarial attacks, cost analysis, and annotator confidence correlation. We further perform a qualitative analysis of our perceived difficulty of the task given the different topics of the passages in our dataset and conclude with recommendations and suggestions that might be of value to people working on future DADC tasks and related annotation interfaces.

1 Introduction

We present quantitative and qualitative analyses of our experience as annotators in the machine reading comprehension shared task for the First Workshop on Dynamic Adversarial Data Collection.[1]. The shared task was a collection of three sub-tasks focused on the selection of excerpts from unstructured texts that best answer a given question (extractive question-answering). The sub-tasks included: (A) the manual creation of question-answer pairs by human annotators, (B) the submission of novel training data (10,000 training examples), and (C) the creation of better extractive question-answering models. In this paper, we focus on our participation in the the manual creation of question-answer pairs task dubbed as "Track 1: Better Annotators".

Machine reading comprehension (MRC) is a type of natural language processing task that relies in the understanding of natural language and knowledge about the world to answer questions about a given text (Rajpurkar et al., 2016). In some cases, state-of-the-art MRC systems are close to or have already started outperforming *standard* human benchmarks (Dzendzik et al., 2021). However, models trained on *standard* datasets (i.e., collected in non-adversarial conditions) do not perform as well when evaluated on adversarially-chosen inputs (Jia and Liang, 2017).

To further challenge models and make them robust against adversarial attacks, researchers have started creating adversarial datasets which continuously change models as they grow stronger. Dynamic Adversarial Data Collection (DADC) is an emergent data collection paradigm explicitly created for the collection of such adversarial datasets. In DADC, human annotators interact with an adversary model or ensemble of models in real-time during the annotation process (Bartolo et al., 2020) to create examples that elicit incorrect predictions from the model (Kaushik et al., 2021). DADC allows for the creation of increasingly more challenging data as well as improved models and benchmarks for adversarial attacks (Dua et al., 2019; Kaushik et al., 2021; Nie et al., 2020).

There is evidence that data collected through adversarial means is distributionally different from standard data. From a lexical point of view, Kaushik et al. (2021) note that "what-" and "how-" questions dominate in adversarial data collection (ADC) as opposed to "who-" and "when-" questions in the standard datasets. In the context of reading comprehension, DADC has been championed by Bartolo et al. (2020), who observe that DADC QA datasets are generally syntactically and lexically more diverse, contain more paraphrases and comparisons, and often require multi-hop inference, especially implicit inference.

*Corresponding author.

[1]https://www.aclweb.org/portal/content/call-participation-shared-task-first-workshop-dynamic-adversarial-data-collection-dadc

Proceedings of the First Workshop on Dynamic Adversarial Data Collection, pages 53 - 60
July 14, 2022 ©2022 Association for Computational Linguistics

Annotation Result	Total	Single Annotator Sessions	Two-Annotator Sessions	Three-Annotator Sessions
Model fooled	45	21	19	5
Model not fooled	43	22	13	8
False negative	10	5	4	1
False positive	2	1	1	0
Total	100	49	37	14

Table 1: Overall annotation results before verification.

Apart from corpus analyses, researchers have also noted certain limitations of the DADC paradigm. For instance, Kiela et al. (2021) note that annotators overfitting on models might lead to cyclical progress and that the dynamically collected data might rely too heavily on the model used, which can potentially be mitigated by mixing in standard data. Similarly, Kaushik et al. (2021) find that DADC models do not respond well to distribution shifts and have problems generalizing to non-DADC tests.

Contributions In this paper, we present our experience as annotators in the reading comprehension shared task for the First Workshop on Dynamic Adversarial Data Collection. Through quantitative and qualitative analyses of a quasi-experimental annotation design, we discuss issues such as cost analysis, annotator confidence, perceived difficulty of the task in relation to the topics of the passages in our dataset, and the issues we encountered while interacting with the system, specifically in relationship with the commonly-used F1 word-overlap metric. We conclude with recommendations and suggestions that might be of value to people working on future DADC tasks and related annotation interfaces.

2 Task Description

Track 1 of the First Workshop on Dynamic Adversarial Data Collection consisted in generating 100 reading comprehension questions from a novel set of annotation passages while competing against the current state-of-the-art QA model (Bartolo et al., 2021), which would remain static throughout the task. Through Dynabench (Kiela et al., 2021),[2] an annotation platform specialized in DADC, annotators would create model-fooling questions that could be answered with a continuous span of text. Successful attacks required the annotators to provide explanations of the question and a hypothesis for the model's failure. These were then subject to a *post hoc* human validation.

2.1 F1 metric and false negatives

During our participation, we discovered two issues with the implementation of the metric used in Dynabench to decide whether the model had been fooled or not.

Dynabench uses a word-overlap metric to calculate the success of the model(s)' responses against those selected by the annotators (Kiela et al., 2021). This metric is calculated as the F1 score of the overlapping words between the answer selected by the annotators and the answer predicted by the model, where model responses with a score above 40% are labeled as a successful answer for the model. For example, the answer "New York" would be considered equivalent to the answer "New York City" (Bartolo et al., 2020).

In practice, we observed that the F1 metric led to many false negatives,[3] or, in other words, to answers that were considered unsuccessful attacks from the annotators when, in reality, the model was wrong. This happened in two different circumstances. First, in the form of incomplete answers where critical information was missing from the model's answer, and the answer was still considered equivalent due to a sufficient word overlap, as in example A from Table 2.

In this case, since "Tinker Tailor Soldier Spy" is a movie, it cannot be said that the first movie and the sequel are equivalent. This behavior was so common that we decided to turn it into an adversarial-attack strategy by forcing the model to provide full answers, which it could not do because of its strong bias towards short answers. For example, we asked questions such as "What is the *full location* of the plot of the TV show?", for which

[2]`https://dynabench.org/`

[3]Notice that, from a model-evaluation perspective, these would be considered false positives.

Question	Model's answer	Annotators' answer
A) What was Eric Fellner working on?	Tinker Tailor Soldier Spy	*A sequel to Tinker Tailor* Soldier Spy
B) At what times is the eastern walkway open to pedestrians only?	5:00 am to 3:30 pm	5:00 am to *6:00 pm, or 9:00 pm during DST*

Table 2: Examples of questions, model answers, and annotators' answers in the data creation procedure. All question-answer examples are adapted from the Dynabench dataset.

the model tended to answer with the bare minimum of information due to being trained using the F1 word-overlap metric.

In other cases, the model selected a different text span than the one selected by the annotators, as in example B from Table 2.

In this case, not only is the model's answer incomplete but *3:30 pm* and *6:00 pm* have entirely different meanings. Cases such as the one above occurred in passages that had two very similar strings in the text. In these cases, the F1 metric lead Dynabench to score in favor of the model even when the answer was incorrect. We believe that the answer provided by the annotators, in cases where annotators are hired as experts in a given domain, should be considered a gold standard subject to the validation process. In other cases, when annotations come from crowdsourcing platforms, the F1 metric could be more adequate.

3 Methodology

Our annotator roster consisted of three annotators with postgraduate degrees in linguistics and natural language processing. One of the annotators spoke English as a first language, while the other two were proficient speakers of English as a second language who completed their graduate degrees in English-speaking universities. For the annotation process, we set up a quasi-experimental design using convenience sampling where approximately half of the annotations would be performed by a single annotator (n=49), and the other half would be performed synchronously by a group of two or more annotators (n=51). Because the annotators live in different time zones, annotator groups did not remain consistent across group sessions.

During the annotation task, the platform randomly picked a passage, usually of the length of a short paragraph (of about 160 words on average) from different topics. Annotators could then choose to create questions for that passage or skip it entirely. Annotators skipped passages when we agreed that it would be difficult to create even a single question to fool the model.[4] Table 1 contains our overall annotation results by the number of annotators. We report our results using the following typology:

Model fooled: Items marked by Dynabench as successful annotations.
Model not fooled: Items marked by Dynabench as unsuccessful annotations.
False negatives: Instances where the model was fooled, but Dynabench marked them as not fooled.[5]
False positives: Items marked by Dynabench as successful annotations but deemed unsuccessful by the annotators.[6]

Even though the limited number of examples does not allow us to draw any strong conclusions about the annotation task, we find our analyses worth presenting as a preliminary step for other annotators to further reflect on the annotation process during the planning stages of any DADC task.[7]

3.1 Model fooled ratio by annotator group

In order to capture if we as annotators are increasingly improving our model-fooling skills, we investigate the progression of the "model fooled / model not fooled" ratio throughout the annotation sessions. Figure 1 summarizes the results.

For the single-annotator group, the progression seems apparent with a progressive fool ratio of 0, .44, .50, and .61. Sessions with two annotators do not have a clear progression (0.57, 0.60, 0.58, and

[4] We did not keep track of the passages we skipped.

[5] Mainly due to F1 score problems.

[6] There are two of these in the dataset and they were products of mistakes the annotators made when selecting the answers on Dynabench that lead to a mismatch between the question asked and the answer given to the model.

[7] The code for our analyses can be found at `https://github.com/fireworks-ai/conference-papers/tree/master/naacl-dadc-2022`

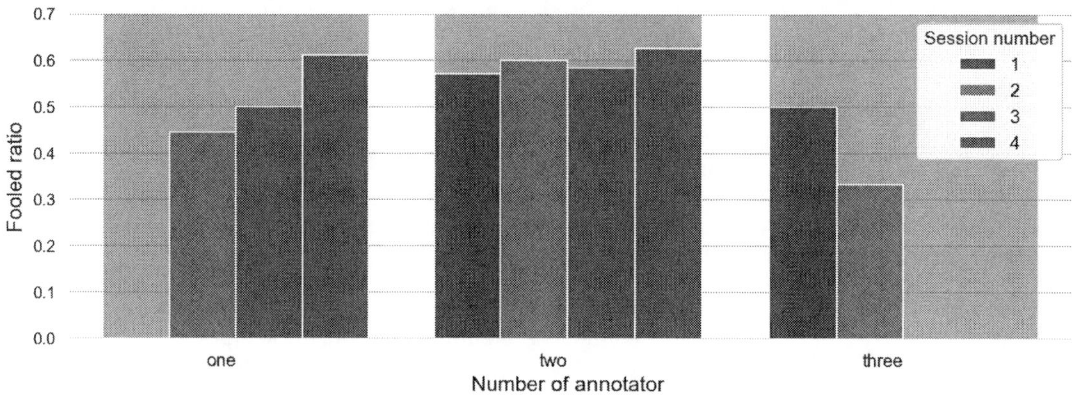

Figure 1: Model fooled ratio by annotator group by session. False negatives and false positives are excluded. Missing sessions had a score of zero.

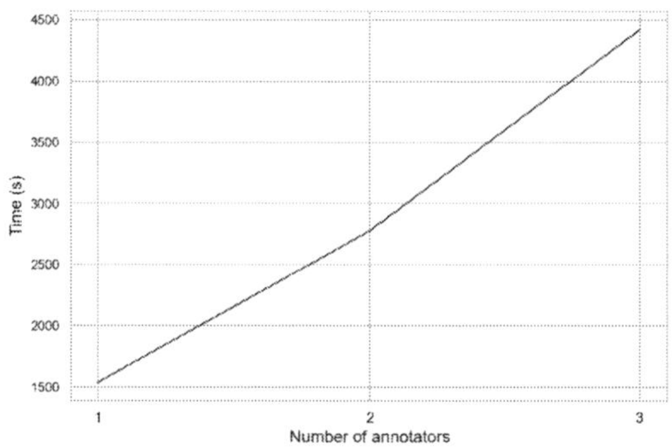

Figure 2: Mean time in seconds spent per annotator for every successful adversarial attack across groups with different annotators.

0.62), which may be because annotators did not remain the same in each session. The worst performance happened with the three-annotator sessions (0.50 and 0.33), which indicates a possible high degree of disagreement across annotators.

3.2 Annotation costs

We investigate the efficiency of the different annotator groups by calculating the mean time per successful adversarial attack. Formally, we define annotation group efficiency $E(g)$ as:

$$E(g) = \frac{\sum_{n}^{k} t_n \times a_n}{N}$$

Where t_n is the total time in seconds spent in annotation session n, k is the total number of sessions for annotator group g, a_n is the number of annotators in the session, and N is the total number of successful adversarial attacks across all the annotation sessions for group g. Table 3 shows annotation efficiency in seconds.

Number of annotators in group	Mean time (s) per successful example
1	**1537.04**
2	2777.58
3	4423.80
Total mean time	8738.42

Table 3: Mean time in seconds spent per annotator for every successful adversarial attack across groups with different annotators.

The single-annotator group took 8h 58' 58" to create 21 model-fooling examples, rendering efficiency of 25' 37" per successful attack. For the group annotations, the two-annotator sessions took 14h 39' 34" to create 19 model-fooling examples, with an efficiency of 46' 17", while the three-annotator sessions took 6h 8' 39" to create five successful examples with an efficiency of 1h 13'. The total time spent on the task was 29h 46" 11'. Figure 2 shows (in seconds) how the time increment is almost linear.

3.3 Confidence scores

Lastly, to better understand why annotation times took longer when working in groups, we investigate the level of confidence agreement between annotators via correlation. To measure confidence agreement, annotators individually logged in confidence scores for all of the 100 questions in our dataset. The scores range between 0 and 3 points, with three being entirely confident that they would fool the model.

We first test our data for normality using the

"normaltest" function of the Python SciPy library (Virtanen et al., 2020). After ensuring that normality tests came out negative across all annotators' ratings ($p < 0.001$), we used the Spearman rank correlation test (Figure 3) as implemented in the Python Pandas library (McKinney, 2010; Reback et al., 2022).

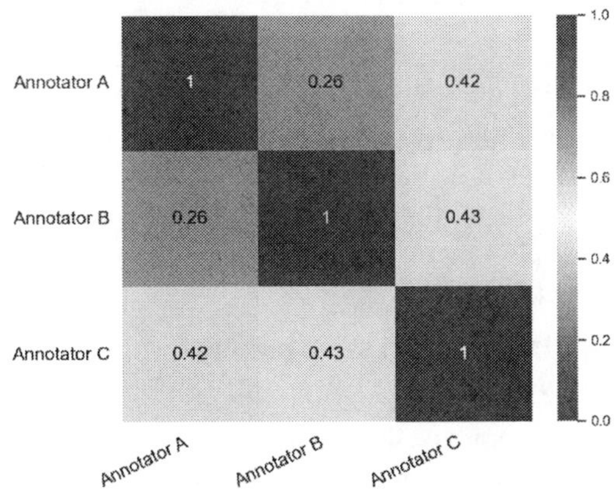

Figure 3: Correlation heatmap of annotators' confidence metrics through the full dataset.

The fact that correlation coefficients range from weak to moderate supports our view that the lower efficiency in annotation costs might be due to differences in how annotators perceive how the model will evaluate their questions. This could lead to more debate during the synchronous annotation sessions. The lack of exponential time increase when more annotators are present, as was the case of the sessions with three annotators, may be due to the fact that annotators were often tired of the feeling of low-productivity of the sessions and were, at times, willing to risk questions without fully debating them.

4 Qualitative Analysis

The relative difficulty of a dynamic adversarial dataset creation task may vary partly as a function of the genre and specific topic of the text passages from which question-answer pairs are drawn. During the shared task, Dynabench randomly assigned passages for the creation of question-answer pairs, revealing several important aspects of this challenge. Topics of the passages used in our data vary as shown by the success by topic scores in Table 4.

Our more successful questions came from music, science, and technology topics. On the one hand, we are more familiar with these topics than comics, sports, and video games. Furthermore, the paragraphs in literature and music tended to be more narrative in nature which, we believe, also made it easier for us to process them and create better questions. Data-heavy, enumeration-based paragraphs typical of sports, history, and TV and movies topics proved more challenging for the creation of model-fooling questions. Still, further examination is necessary to understand each of these possibilities separately.

A closer examination of the DADC task included evaluating the success of different strategies for creating questions. Overall, the model successfully answered questions about dates and names, as well as questions that could be answered with a single short phrase, especially if that phrase was produced as an appositive. For example, asking "Which political associate of Abraham Lincoln was aware of his illness while traveling from Washington DC to Gettysburg?" allowed the model to select a name as the answer, which it did with a high degree of success, even when multiple distractor names appeared in the same paragraph. On the other hand, formulating questions that required longer answers, especially questions that asked for both "what" and "why", frequently fooled the model. Furthermore, requiring references to multiple non-contiguous portions of the passage to make predictions also often fooled the model. Still, using synonymous words or phrases or similar sentence structures to the critical portions of the passage allowed the model to make correct predictions, even when these other strategies may have fooled it under different circumstances.

5 Discussion

Based on the experience with DADC shared task Track 1, we recommend several strategies to improve the efficiency of data collection.

5.1 Experimenting with the task

We found that allowing annotators to run "dry" trials before starting data collection, as done by the organizers of the DADC Shared Task, might help them form initial hypotheses about the potential weaknesses of the model and what strategies could be helpful to fool it, e.g., targeting different capabilities such as NER or coreference resolution. Additionally, it could be possible that once annotators are familiarized with the task and understand what examples have a better chance of fooling the

	Comics	History	Literature	Music	Science	Sports	TV and Movies	Technology	Video Games
Model Fooled	1	6	5	5	4	8	9	2	5
Total Items	3	16	7	9	4	23	20	4	14
Ratio Fooled	0.33	0.37	**0.71**	**0.55**	**1.00**	0.35	0.45	0.50	0.36

Table 4: Number of times our questions fooled the model out of the total number of questions we generated for each passage topic in our dataset. False negatives and false positives are included in the total number of items.

model, productivity between multiple annotators might increase as their confidence starts to align.

5.2 Familiarity with the domain

We believe it may be significantly easier to come up with good-quality questions if the annotators are familiar with the domain of the contexts. Not only can they read and understand the paragraphs faster, but it is easier to abstract from the immediate context and, thus, ask more challenging questions. Annotation managers of campaigns with heterogeneous datasets might want to consider recruiting experts for technical or specific sub-domains and crowdsourced workers for those texts consisting of general knowledge.

5.3 Having a list of strategies

Keeping a rough track of what annotation strategies worked best proved useful to us during annotation. As an example of the types of strategies that annotators can keep track of and implement, below we list the strategies we favored for creating model-fooling questions.

1. Play with the pragmatics of the question, for instance:

   Question: What is the full location of the plot of this TV show?

   Annotators' answer: A mysterious island somewhere in the South Pacific Ocean

   Model's answer: South Pacific Ocean

   Explanation: The model is biased towards the shortest answer, which does not always cover the information human need as an answer (Grice's principle of quantity)

2. Change the register, e.g., ask a question as a five-year-old would.
3. Whenever possible, ask a question that requires a holistic understanding of the whole paragraph (not just a particular sentence).
4. Ask questions that require common sense reasoning, e.g., about the causes and effects of events.
5. Ask questions about entities that appear multiple times or have multiple instances in the paragraph.

5.4 Discussing created prompts with other annotators

Another practice that can help is to work in teams whereby annotators would come up with questions in isolation and then rank and further modify them in a brainstorming session. In our experience, having two annotators in one session was almost as efficient as having only one annotator and made the task more engaging,ludic and, consequently, less tedious, potentially reducing the risk of burnout syndrome.

5.5 Suggestions for future DADC annotation interfaces

Because DADC annotation applies to NLI and QA datasets (Kiela et al., 2021), we believe that specific considerations would be necessary for future projects that make use of a dedicated DADC interface, including the following:

- Given that one of the issues we observed was that many of the successful questions were unnatural and, thus, probably, not helpful for real-life scenarios, annotation platforms could include a naturality score to encourage annotators to create data that will be used in real-world scenarios.
- Because the word-overlap F1 threshold seems to vary depending on what is enough information and the appropriate information needed to answer specific questions, we believe that a language model could be trained to replace or aid the F1 metric.
- Annotation interfaces could also help annotators by displaying relevant visualizations of

the training data so that annotators could try to fool the model in those cases where the model contains little or no data. For example, Bartolo et al. (2020, pp. 17-19) provide bar plots and sunburst plots[8] of question types and answer types for each of their modified datasets. We believe that displaying such visualizations to the annotators in a targeted way could potentially increase their performance while also helping balance the creation of datasets.

- Finally, we believe that augmenting the interface with functionality for storing and managing annotation strategies such as the ones mentioned above, together with their rate of effectiveness, could make the annotation process more efficient.

5.6 Final considerations

Beyond any of the suggestions above, we believe that the DADC has certain limitations that annotation campaigns should be aware of.

In our experience in the context of this extractive QA task, we found it extremely difficult to fool the model, primarily because of its powerful lexical and syntactic reasoning capabilities. This was partly because we were constrained to create questions that a continuous string of text could answer. In many cases, we relied on very complex lexical and syntactic inferences (e.g., violating syntactic islands), which often led to unnatural questions that were unlikely to appear in the real world.

The problem of creating model-fooling examples has already been acknowledged in previous research (Bartolo et al., 2020; Kiela et al., 2021) and is generally addressed by either providing question templates to edit or mixing questions from other "more naturally-distributed" datasets. We want to draw the attention of anyone wishing to apply DADC to their problem of this risk.

Kiela et al. (2021) note that applying DADC for generative QA is not a straightforward task. However, it is perhaps in generative tasks where DADC could offer more value. Given how powerful the SOTA models are, the DADC extractive datasets seem doomed to be eventually skewed towards long and unnatural examples. This is one of ours: "Despite knowledge of which fact does Buffy still allow herself to pass at the hands of an enemy, protecting the one to whom the fact relates by doing so?"

[8]The dataset statistics are only available in the pre-print version of their paper, available at: https://arxiv.org/abs/2002.00293

Acknowledgements

We thank the organizers and sponsors of the first DADC shared task, especially Max Bartolo, who was in direct contact with us and provided us with the data we needed for our analyses. We would also like to thank Dr. Anders Søgaard for his valuable insights during the revision of this article.

References

Max Bartolo, Alastair Roberts, Johannes Welbl, Sebastian Riedel, and Pontus Stenetorp. 2020. Beat the AI: Investigating Adversarial Human Annotation for Reading Comprehension. *Transactions of the Association for Computational Linguistics*, 8:662–678.

Max Bartolo, Tristan Thrush, Robin Jia, Sebastian Riedel, Pontus Stenetorp, and Douwe Kiela. 2021. Improving question answering model robustness with synthetic adversarial data generation. In *Proceedings of the 2021 Conference on Empirical Methods in Natural Language Processing*, pages 8830–8848, Online and Punta Cana, Dominican Republic. Association for Computational Linguistics.

Dheeru Dua, Yizhong Wang, Pradeep Dasigi, Gabriel Stanovsky, Sameer Singh, and Matt Gardner. 2019. DROP: A reading comprehension benchmark requiring discrete reasoning over paragraphs. In *Proceedings of the 2019 Conference of the North American Chapter of the Association for Computational Linguistics: Human Language Technologies, Volume 1 (Long and Short Papers)*, pages 2368–2378, Minneapolis, Minnesota. Association for Computational Linguistics.

Daria Dzendzik, Jennifer Foster, and Carl Vogel. 2021. English machine reading comprehension datasets: A survey. In *Proceedings of the 2021 Conference on Empirical Methods in Natural Language Processing*, pages 8784–8804, Online and Punta Cana, Dominican Republic. Association for Computational Linguistics.

Robin Jia and Percy Liang. 2017. Adversarial examples for evaluating reading comprehension systems. In *Proceedings of the 2017 Conference on Empirical Methods in Natural Language Processing*, pages 2021–2031, Copenhagen, Denmark. Association for Computational Linguistics.

Divyansh Kaushik, Douwe Kiela, Zachary C. Lipton, and Wen-tau Yih. 2021. On the efficacy of adversarial data collection for question answering: Results from a large-scale randomized study. In *Proceedings of the 59th Annual Meeting of the Association for Computational Linguistics and the 11th International Joint Conference on Natural Language Processing*

(Volume 1: Long Papers), pages 6618–6633, Online. Association for Computational Linguistics.

Douwe Kiela, Max Bartolo, Yixin Nie, Divyansh Kaushik, Atticus Geiger, Zhengxuan Wu, Bertie Vidgen, Grusha Prasad, Amanpreet Singh, Pratik Ringshia, Zhiyi Ma, Tristan Thrush, Sebastian Riedel, Zeerak Waseem, Pontus Stenetorp, Robin Jia, Mohit Bansal, Christopher Potts, and Adina Williams. 2021. Dynabench: Rethinking benchmarking in NLP. In *Proceedings of the 2021 Conference of the North American Chapter of the Association for Computational Linguistics: Human Language Technologies*, pages 4110–4124, Online. Association for Computational Linguistics.

Wes McKinney. 2010. Data Structures for Statistical Computing in Python. In *Proceedings of the 9th Python in Science Conference*, pages 56 – 61.

Yixin Nie, Adina Williams, Emily Dinan, Mohit Bansal, Jason Weston, and Douwe Kiela. 2020. Adversarial NLI: A new benchmark for natural language understanding. In *Proceedings of the 58th Annual Meeting of the Association for Computational Linguistics*, pages 4885–4901, Online. Association for Computational Linguistics.

Pranav Rajpurkar, Jian Zhang, Konstantin Lopyrev, and Percy Liang. 2016. SQuAD: 100,000+ questions for machine comprehension of text. In *Proceedings of the 2016 Conference on Empirical Methods in Natural Language Processing*, pages 2383–2392, Austin, Texas. Association for Computational Linguistics.

Jeff Reback, jbrockmendel, Wes McKinney, Joris Van den Bossche, Tom Augspurger, Matthew Roeschke, Simon Hawkins, Phillip Cloud, gfyoung, Sinhrks, Patrick Hoefler, Adam Klein, Terji Petersen, Jeff Tratner, Chang She, William Ayd, Shahar Naveh, JHM Darbyshire, Marc Garcia, Richard Shadrach, Jeremy Schendel, Andy Hayden, Daniel Saxton, Marco Edward Gorelli, Fangchen Li, Matthew Zeitlin, Vytautas Jancauskas, Ali McMaster, Torsten Wörtwein, and Pietro Battiston. 2022. pandas-dev/pandas: Pandas 1.4.2.

Pauli Virtanen, Ralf Gommers, Travis E. Oliphant, Matt Haberland, Tyler Reddy, David Cournapeau, Evgeni Burovski, Pearu Peterson, Warren Weckesser, Jonathan Bright, Stéfan J. van der Walt, Matthew Brett, Joshua Wilson, K. Jarrod Millman, Nikolay Mayorov, Andrew R. J. Nelson, Eric Jones, Robert Kern, Eric Larson, C J Carey, İlhan Polat, Yu Feng, Eric W. Moore, Jake VanderPlas, Denis Laxalde, Josef Perktold, Robert Cimrman, Ian Henriksen, E. A. Quintero, Charles R. Harris, Anne M. Archibald, Antônio H. Ribeiro, Fabian Pedregosa, Paul van Mulbregt, and SciPy 1.0 Contributors. 2020. SciPy 1.0: fundamental algorithms for scientific computing in Python. *Nature Methods*, 17(3):261–272.

Generalized Quantifiers as a Source of Error in Multilingual NLU Benchmarks

Ruixiang Cui, Daniel Hershcovich, Anders Søgaard
University of Copenhagen
{rc, dh, soegaard}@di.ku.dk

Abstract

Logical approaches to representing language have developed and evaluated computational models of quantifier words since the 19th century, but today's NLU models still struggle to capture their semantics. We rely on Generalized Quantifier Theory for language-independent representations of the semantics of quantifier words, to quantify their contribution to the errors of NLU models. We find that quantifiers are pervasive in NLU benchmarks, and their occurrence at test time is associated with performance drops. Multilingual models also exhibit unsatisfying quantifier reasoning abilities, but not necessarily worse for non-English languages. To facilitate directly-targeted probing, we present an adversarial generalized quantifier NLI task (GQNLI) and show that pre-trained language models have a clear lack of robustness in generalized quantifier reasoning.

Proceedings of the First Workshop on Dynamic Adversarial Data Collection, pages 61 - 61
July 14, 2022 ©2022 Association for Computational Linguistics

Adversarially Constructed Evaluation Sets Are More Challenging, but May Not Be Fair

Jason Phang,[1] **Angelica Chen,**[1] **William Huang,**[2*] **Samuel R. Bowman**[134]
[1]Center for Data Science, New York University
[2]Capital One
[3]Dept. of Linguistics, New York University
[4]Dept. of Computer Science, New York University

Correspondence: jasonphang@nyu.edu

Abstract

Large language models increasingly saturate existing task benchmarks, in some cases outperforming humans, leaving little headroom with which to measure further progress. Adversarial dataset creation, which builds datasets using examples that a target system outputs incorrect predictions for, has been proposed as a strategy to construct more challenging datasets, avoiding the more serious challenge of building more precise benchmarks by conventional means. In this work, we study the impact of applying three common approaches for adversarial dataset creation: (1) filtering out easy examples (AFLite), (2) perturbing examples (TextFooler), and (3) model-in-the-loop data collection (ANLI and AdversarialQA), across 18 different adversary models. We find that all three methods can produce more challenging datasets, with stronger adversary models lowering the performance of evaluated models more. However, the resulting ranking of the evaluated models can also be unstable and highly sensitive to the choice of adversary model. Moreover, we find that AFLite oversamples examples with low annotator agreement, meaning that model comparisons hinge on the examples that are most contentious for humans. We recommend that researchers tread carefully when using adversarial methods for building evaluation datasets.

*Work done while at NYU.

Proceedings of the First Workshop on Dynamic Adversarial Data Collection, pages 62 - 62
July 14, 2022

Author Index